Emrah Celik

Additive Manufacturing

Science and Technology

2nd Edition

DE GRUYTER

T0356899

Author

Dr. Emrah Celik
Department of Mechanical and Aerospace Engineering
University of Miami
1251 Memorial Drive
Coral Gables, FL 33146
USA
e.celik@miami.edu

ISBN 978-1-5015-2023-5
e-ISBN (PDF) 978-1-5015-2024-2
e-ISBN (EPUB) 978-1-5015-1722-8

Library of Congress Control Number: 2025931340

Bibliographic information published by the Deutsche Nationalbibliothek
The Deutsche Nationalbibliothek lists this publication in the Deutsche Nationalbibliografie;
detailed bibliographic data are available on the Internet at http://dnb.dnb.de.

© 2025 Walter de Gruyter GmbH, Berlin/Boston, Genthiner Straße 13, 10785 Berlin
Cover image: michal-rojek/iStock/Getty Images Plus
Typesetting: Integra Software Services Pvt. Ltd.

www.degruyter.com
Questions about General Product Safety Regulation:
productsafety@degruyterbrill.com

Emrah Celik

Additive Manufacturing

Also of interest

Additive Manufacturing of Structural Electronics
Marcin Słoma, 2024
ISBN 978-3-11-079359-8, e-ISBN (PDF) 978-3-11-079360-4

3D Printing.
Polymer, Metal and Gel Based Additive Manufacturing
Edited by: Rupinder Singh, Ranvijay Kumar, Vinay Kumar and
J. Paulo Davim, 2024
ISBN 978-3-11-108061-1; e-ISBN (PDF) 978-3-11-108163-2

3D Printing Technologies.
Digital Manufacturing, Artificial Intelligence, Industry 4.0
Edited by: Ajay Kumar, Parveen Kumar, Naveen Sharma and
Ashish Kumar Srivastava, 2024
ISBN: 978-3-11-121459-7; e-ISBN (PDF) 978-3-11-121511-2

Additive and Subtractive Manufacturing.
Emergent Technologies
Edited by: J. Paulo Davim, 2022
ISBN 978-3-11-077677-5, e-ISBN (PDF) 978-3-11-054977-5

Bulk Metallic Glasses and Their Composites.
Additive Manufacturing and Modeling and Simulation
Muhammad Musaddique Ali Rafique, 2021
ISBN 978-3-11-074721-8, e-ISBN (PDF) 978-3-11-074723-2

To mom and dad

Preface

Additive manufacturing, commonly known as 3D printing, is a disruptive technology that creates engineering materials layer by layer, based on a predesigned and tessellated computer-aided design model. This method enables more cost-effective and flexible production by reducing the need for large inventories of raw materials, minimizing labor costs, and allowing printers to be easily transported to various locations. Unlike traditional manufacturing processes that remove material, 3D printing's additive approach results in minimal production waste. Today, with the growing accessibility of 3D printers worldwide, this technology is rapidly evolving into a convenient, home-based manufacturing solution.

Despite its distinct advantages over traditional manufacturing, additive manufacturing still faces notable challenges, including slower production speeds, a limited range of available materials, high equipment costs, poor interface bonding, reduced fatigue performance of printed materials, and low reproducibility of material properties. These issues, which need to be fully addressed or mitigated, are thoroughly examined throughout the book, along with potential solutions. The book also explores future trends in additive manufacturing from the author's perspective, highlighting how the technology will continue to evolve and further impact our lives.

The goal of this book is to provide students, researchers, and practicing engineers with a deeper understanding of current additive manufacturing technologies. It offers a comprehensive exploration of both the theoretical foundations and practical aspects of various additive manufacturing methods. By combining fundamental science with the latest advancements, the book aims to help readers grasp essential concepts while keeping them informed about the most recent developments in each area of manufacturing.

The structure of this book is designed to serve both as an educational resource for individual learners and as a textbook for additive manufacturing courses. The first chapter introduces the concept of additive manufacturing, outlining its seven major categories. It also highlights the advantages of this technology, along with the limitations that prevent it from fully replacing conventional manufacturing methods. Additionally, the chapter provides a brief history of additive manufacturing, tracing its evolution to give readers insights into its development pace and trends, helping them anticipate future advancements in the field.

The remaining chapters are organized by the types of materials used in additive manufacturing. As polymers are the most widely used, the book begins with Chapter 2, which covers the additive manufacturing of thermoplastic, thermosetting, and elastomeric polymers. Chapters 3, 4, and 5 then explore the additive manufacturing of other key engineering materials: polymer composites, metals, and ceramics, respectively. Each material system is compared in terms of manufacturability, cost, mechanical performance, and thermal resistance. These chapters also provide a detailed overview of the material options and manufacturing technologies for each category, offering insight into the reasons behind material selection for different applications.

https://doi.org/10.1515/9781501520242-202

Chapter 6 investigates the use of additive manufacturing on tissue engineering applications. A specialized version of additive manufacturing known as bioprinting is described in detail. The chapter explores the transformative potential of bioprinting, its use across various tissues and organs, and highlights current success stories. It also evaluates how the choice of feedstock materials known as bioink affects the quality of bioprinting and provides an overview of the bioink materials currently in use. Alongside discussing the benefits and recent advancements in bioprinting, the chapter addresses the significant challenges and limitations that the field still faces.

The true advantage of additive manufacturing lies in its ability to produce complex structures with enhanced performance and/or reduced weight compared to traditional unoptimized parts. In other words, when component topology is optimized to improve functionality, additive manufacturing can bring these intricate designs to life, regardless of their complexity. Chapter 7 covers the principles of topology optimization, its benefits, and how it is integrated into the world of additive manufacturing.

Chapter 8 explores advanced concepts in additive manufacturing, which hold significant potential to revolutionize manufacturing technologies. These concepts include additive manufacturing of unconventional materials, such as thermoelectric systems, which convert heat into electricity. Given the vast amounts of waste heat generated, the ability to 3D print these systems could have a profound impact on daily life. The chapter also delves into hybrid manufacturing, which combines different materials or manufacturing systems in unique ways. Additionally, it introduces 4D printing, a smart material manufacturing technique in which additively manufactured structures undergo time-dependent transformations in a controlled manner. Finally, the chapter reviews the integration of artificial intelligence on additive manufacturing science and technology, and the book concludes by discussing how these cutting-edge technologies pave the way for future advancements, adding new dimensions and functionalities to additive manufacturing systems.

Additive manufacturing is a technology that is revolutionizing the design and manufacturing process. Its impact is expected to grow in the near future as existing technologies advance, manufacturing costs decrease, and new concepts like smart manufacturing and artificial intelligence are integrated into material development and processing. This book covers the theoretical concepts, latest technologies, and future trends in additive manufacturing. It aims to assist students and individuals looking to contribute to the field, whether as users or developers – both of whom are essential for the ongoing advancement of this technology.

Contents

1 Introduction

1.1 A disruptive technology, additive manufacturing

Additive manufacturing (AM) is a revolutionary technology that constructs three-dimensional (3D) objects by adding successive layers of a material. This technique can be used to fabricate a variety of materials, including polymers, metals, ceramics, composites, and biological materials. Because AM enables the creation of complex geometries without the need for tooling, its early applications primarily focused on rapid prototyping for visualization models. However, thanks to significant advancements in material libraries and the quality of fabricated parts, AM is now increasingly utilized to produce end products across various industries, including aerospace, dentistry, medical implants, automotive, and even fashion design.

Unlike traditional subtractive manufacturing, which involves removing materials to create objects, AM builds objects layer by layer by joining materials. Over the past century, subtractive manufacturing has significantly influenced fabrication and prototyping since its introduction. However, the manufacturing industry is now on the brink of a new revolution, thanks to the innovative design and fabrication possibilities offered by AM. Figure 1.1 illustrates the key differences between subtractive manufacturing and additive manufacturing technologies. As the name suggests, subtractive manufacturing removes material through machining or cutting, which can be done manually or using computer numerical control (CNC) machining. In contrast, AM is also a computer-controlled process, but it adds successive layers of material to create a 3D object.

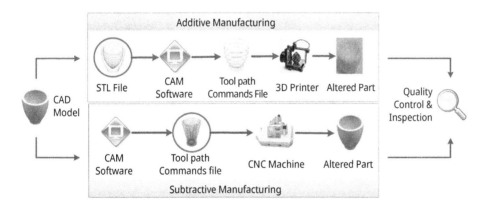

Fig. 1.1: Additive versus subtractive manufacturing. Figure was reprinted from [1].

Traditional subtractive manufacturing generally offers advantages such as lower cost per part (especially for large-scale production), high speed, good component strength, and a wide range of material options. However, emerging AM technologies

https://doi.org/10.1515/9781501520242-001

are rapidly bridging these gaps, introducing a new design space that includes on-demand fabrication, customization, and complex geometries, thereby redefining contemporary manufacturing practices.

As discussed in the next section, numerous AM technologies have been developed for different manufacturing purposes. Despite their differences, all AM methods share common processing steps, as illustrated in Fig. 1.2. The AM process begins with designing the component using 3D modeling software, such as computer-aided design (CAD). Alternatively, the model can be created through reverse engineering using a 3D scanner or photography. Once the computer model is ready, it is transferred to the slicing software, which converts the 3D model into a series of 2D sections. The AM equipment then interprets this data and lays down successive layers of liquid, powder, sheet material, or other forms to fabricate the 3D object. The minimum layer thickness determines the manufacturing quality, which varies depending on the selected machine and the AM process. Once the AM is complete, the object is removed from the AM device, and optional postprocessing steps – such as cleaning, sanding, coating, painting, compacting, or heat treatment – can be applied to enhance performance or improve the aesthetic appearance of the component. Postprocessing may require additional machines and tools.

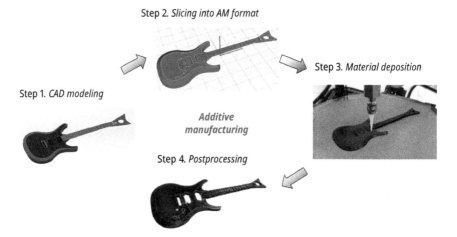

Fig. 1.2: Main processing steps of additive manufacturing: CAD modeling, AM format slicing, material deposition, and postprocessing.

1.2 Advantages of additive manufacturing over traditional manufacturing

AM has unique applications across a wide range of fields, including electronics, aerospace, automotive engineering, and even fashion design, as illustrated in Fig. 1.3. This versatility is attributed to the advantages of AM compared to traditional manufacturing methods. These benefits are summarized below:

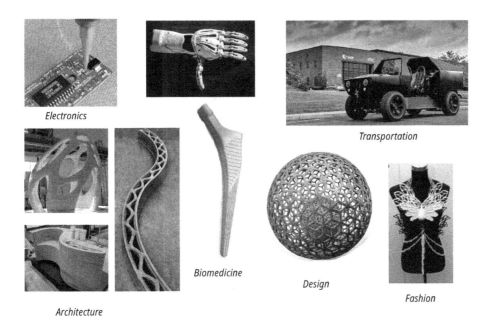

Electronics

Transportation

Architecture

Biomedicine

Design

Fashion

Fig. 1.3: Various applications of additive manufacturing: concrete printing (architecture) [2], transportation [3], and fashion [4] application. Images were reprinted with permission.

Enhanced design capability: AM systems enable the production of moving parts (such as hinges and chains) in a single print process, as well as the creation of complex structures that are impossible to achieve with traditional manufacturing methods. Additionally, AM removes traditional manufacturing constraints, allowing for greater design flexibility and a reduction in the number of parts needed. The redesign process can be carried out digitally using CAD software, enabling rapid fabrication of new components. Because AM is user-friendly and not labor-intensive, it opens up opportunities for a diverse range of creators – not just engineers but also architects, artists, and even students – to utilize this technology in their projects. As a result, AM technologies have fostered a new design space and generated significant interest within the community.

No tooling: In AM, unlike many traditional techniques, there is no need for jigs, fixtures, or molds to secure and shape the parts being produced. Components are fabricated directly on a print bed or on an already-fabricated part, resulting in significant cost savings compared to the expensive tooling required for traditional manufacturing systems.

On-demand manufacturing: Parts can be produced as soon as the CAD model is prepared, allowing for on-demand fabrication. These CAD models can be sent digitally, enabling manufacturing to take place at home or in remote locations near consumers, thereby eliminating the need for transportation of finished parts.

Rapid prototyping: AM allows for quick production of models, taking hours rather than weeks compared to conventional techniques that have more geometric limitations. This rapid prototyping enables makers and designers to test prototypes swiftly, saving significant time in the design process and facilitating faster development of the final product.

Customization: Each part can be uniquely fabricated through the AM process without incurring additional fabrication costs. This capability is particularly beneficial in biomedical applications, where parts can be fully tailored to meet the specific needs of different patients.

Minimal material waste: Since AM builds 3D objects by adding material rather than removing it, it uses nearly the exact amount of material needed for production, resulting in little to no waste. Any support material or excess powder generated during the process can often be recycled and reused for subsequent production, which reduces material costs associated with waste and waste removal. This also lessens the environmental impact associated with waste in traditional manufacturing.

Low cost for small production runs: With advancements in technology, AM systems are becoming more affordable and portable compared to traditional manufacturing methods. The low initial investment required for AM machines is attracting increased interest in these systems. The cost of fabrication is highly dependent on the number of parts produced and their complexity. While traditional manufacturing is typically more advantageous for producing large quantities of geometrically simple parts, AM becomes highly competitive or even cheaper for small production runs of complex parts.

1.3 Classification of additive manufacturing technologies

Numerous AM technologies are employed to fabricate a variety of materials. In 2010, the American Society for Testing and Materials (ASTM) established a set of standards that categorize AM processes into seven major groups, as illustrated in Fig. 1.4 [5]. Each of these seven methods differs significantly in its approach to layer-by-layer manufacturing. The following sections provide an explanation of each category.

1.3.1 Vat polymerization

Vat polymerization, also known as vat photopolymerization, is a 3D printing technology that involves selectively curing a liquid photopolymer within a vat using a light

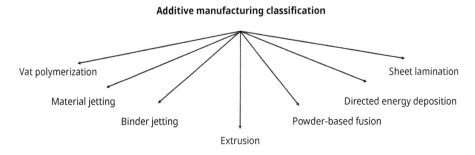

Fig. 1.4: ASTM classification of the seven additive manufacturing technologies.

source. The light solidifies the resin in specific areas while a platform moves the built object downward (or upward) after each layer is cured. This process continues, layer by layer, until the 3D physical object is fully formed. Once completed, the resin in the vat is drained, and the object is removed. Typically, photopolymerized samples undergo postcuring under ultraviolet (UV) light to ensure complete curing and achieve maximum strength.

Most photocurable resins are composed of mixtures of monomers combined with oligomers (short chains of monomers) and photoinitiators. As depicted in Fig. 1.5, oligomers and monomers remain separate in the uncured liquid photopolymer resins. When UV light is applied, the photoinitiators are activated, causing the monomer and oligomer units to cross-link. This chemical process, known as photopolymerization or photocuring, converts liquid photopolymers into solid components.

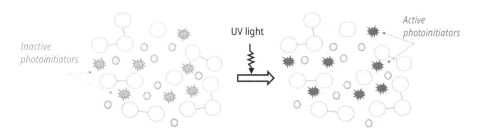

Fig. 1.5: Schematic of photopolymerization process.

Photopolymerization processes can be categorized into two main types based on the photoinitiator and polymerization mechanism: (1) free radical polymerization and (2) ionic photopolymerization. Free radical photopolymerization is commonly employed in AM technologies such as vat polymerization and material jetting. This process occurs in several stages: first, the photoinitiator is activated by exposure to radiation within a suitable wavelength range. Next, free radicals are generated through a reaction between the photoinitiator and monomer molecules. These free radicals then propagate by forming long polymer chains, leading to cross-linking. In the final stage, the polymerization process terminates, typically through one of three mechanisms: recombination (where two chains combine), disproportionation (where one radical cancels out another without combining), or occlusion (where free radicals become trapped within the polymer network) [6]. UV-curable resins, particularly those based on acrylates, are the most widely used in these processes due to their high reactivity and short curing times (often fractions of a second). These resins are available in a variety of monomer and oligomer types [7].

While free radical polymerization is the most widely used photopolymerization method, ionic curing systems are gaining popularity in AM applications. Ionic photopolymerization follows similar steps to free radical polymerization: photoinitiator activation by UV light, propagation, and termination. However, in ionic curing, reactive ions – rather than free radicals – serve as the cross-linking agents for monomers and oligomers. Termination occurs when the ion is neutralized or stabilized. Ionic photopolymerization offers several advantages over free radical processes, including resistance to oxygen inhibition, reduced sensitivity to water, and the ability to polymerize monomers such as vinyl ethers, epoxides, and other heterocyclic compounds that cannot undergo free radical polymerization [7].

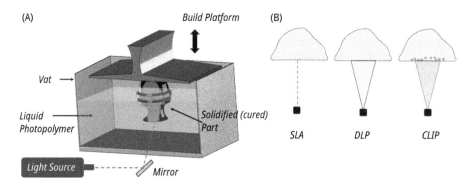

Fig. 1.6: (A) Schematic of part fabrication in vat polymerization technology and (B) different types of vat polymerization process (SLA, DLP, and CLIP).

Vat polymerization technologies are classified into four categories, with the main distinction based on the type of light source and curing process used for the polymer material, even though the core concept of polymerization remains consistent. The operational principles of the three major vat polymerization techniques are illustrated in Fig. 1.6.

1.3.1.1 Stereolithography (SLA)
Stereolithography (SLA) is the oldest AM technology and is widely adopted today. In this process, a focused UV laser beam is directed onto a photosensitive resin using

motor-controlled mirrors. When the light strikes the liquid resin, a chemical reaction occurs, curing the resin and forming a solid layer of the desired 3D object.

1.3.1.2 Digital light processing (DLP)

Digital light processing (DLP) differs from SLA in that the light is projected onto the liquid polymer as a 2D image rather than using a focused beam to raster across the surface. A digital projector reflects the entire image onto the resin, curing the entire 2D layer at once. This results in faster print times compared to SLA, as each layer is solidified simultaneously. A recent modification of DLP replaces the projector with an LCD screen, which acts as a mask for UV light emitted from an LED array. The UV light passes through the LCD screen, curing the entire 2D layer at once, similar to a projector. This innovation has significantly reduced the cost and size of DLP systems, while maintaining faster speeds compared to SLA.

1.3.1.3 Continuous liquid interface production (CLIP)

Continuous liquid interface production (CLIP) is a relatively recent technology introduced by Carbon3D in 2015, offering a novel approach to speed up the 3D printing process [8]. In traditional SLA and DLP processes, solidification occurs at the bottom of a vat with a transparent window. To prevent the resin from sticking to the window, the build platform must move up and down, creating suction that breaks the adhesion between the object and the window. CLIP technology eliminates this issue by using an oxygen-permeable membrane at the bottom of the vat to inhibit resin solidification in a specific zone near the clear window. In the CLIP process, a UV light shines through a transparent portion of the vat bottom using an LCD screen. As the object is slowly lifted, resin flows underneath and remains in contact with the object. The oxygen-permeable membrane below the resin creates a "dead zone" where photopolymerization is inhibited, preventing the resin from adhering to the bottom plate. Unlike standard SLA and DLP, CLIP enables continuous printing, making it significantly faster. This continuous process makes CLIP one of the fastest AM methods. However, like other vat polymerization techniques, postprocessing is required to clean the printed part and fully cure it through additional postcuring steps.

1.3.1.4 Volumetric vat manufacturing

In 2017, volumetric vat AM was introduced, offering a new approach to photopolymerization by creating objects in 3D rather than through the traditional layer-by-layer method [9]. This technique is similar to computed tomography (CT), where a series of X-ray scans are taken from different angles, and these 2D images are processed using computer algorithms to reconstruct a 3D image. Unlike CT imaging, in volumetric vat 3D printing, the process begins with a 3D CAD model, which is then converted to 2D projections from various orientations using tomography algorithms. When these projections are directed into a homogeneous volume of photopolymer material, the cumulative light dose absorbed recreates the shape of the 3D object within the material [10]. Areas receiving higher light doses solidify, while regions with lower doses remain unsolidified. By projecting light throughout the 3D volume simultaneously at varying intensities, the photopolymerization process is significantly accelerated compared to layer-by-layer techniques. While still in the developmental stage, this technique promises faster printing speeds, and improvements in light control and photopolymerization precision are expected in the near future.

Vat polymerization offers several advantages over other AM techniques, including the ability to fabricate parts with extremely high resolution, even down to the nanoscale. It can also produce large parts by utilizing vats with significant volumes, and near-transparent objects can be created – something difficult to achieve with other AM methods, where the interface between layers causes high light diffraction. However, this technique has some limitations. It is compatible with only a limited range of UV-curable resins, which are generally not very durable, strong, or stable materials. Additionally, postcuring under UV light is often necessary to fully solidify the printed object. Some photocurable resins can pose health risks, requiring special gloves and ventilation during printing, and postprocessing until the resin is fully cured. Depending on the complexity of the design, support structures may be required, which increases the material waste and extends the fabrication time.

1.3.2 Material jetting

Material jetting is a 3D printing technique similar to the standard inkjet document printing, but instead of spraying ink onto paper, it jets photopolymer or wax materials onto a build platform. As droplets of the polymer resin are deposited layer by layer, they are solidified using UV light until the full 3D object is formed. This process also enables the fabrication of multiple materials within a single object. Figure 1.7 illustrates the material jetting process and provides examples of parts produced with this technology.

Material jetting processes require support, which is often 3D printed simultaneously from a dissolvable material. The support material is then removed during the postprocessing step. Depending on the type of the support material used and the support removal technique, there are two patented technologies: PolyJet printing and multijet printing (MJP) used by Stratasys and 3D Systems. In PolyJet technology, the support material is a combination of propylene, acrylic monomer, polyethylene, and glycerin [11]. To remove the support material, pressurized water is sprayed over the part, and the remaining support material is removed chemically by dipping the part into chemical solvent. On the other hand, MJP technology uses meltable paraffin wax as the support structure. To remove the wax support, the printed sample is heated in an oven over melting temperature of the wax followed by wiping out of the wax material out of the sample.

Material jetting 3D printing technology is an excellent choice for creating realistic prototypes, offering high levels of detail, accuracy, and smooth surface finishes. It allows designers to print in multiple colors and use a variety of materials in a single print. These systems provide a broad selection of materials and combinations, ranging from rigid to rubber-like and opaque to transparent. However, the main drawbacks of material jetting are the high costs associated with the printing equipment and UV-activated photopolymers, as well as the gradual degradation of these materials' mechanical properties over time.

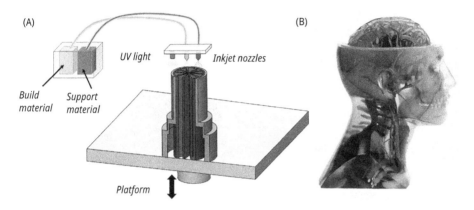

Fig. 1.7: (A) Schematic of material jetting technology and (B) fully colored parts fabricated by a material jetting system. Image was reprinted with permission from Stratasys.

1.3.3 Binder jetting

Binder jetting is an AM process in which a binding material is sprayed (or jetted) over powder particles to bond them together, layer by layer, to create a 3D object. A schematic of this process is illustrated in Fig. 1.8. Common materials used in binder jetting include metals and ceramics in powder form. During the process, inkjet nozzles – similar to those found in standard paper inkjet printers – spray droplets of a binding material onto the powder print bed, bonding the powder particles in those areas. Once a layer is completed, the build platform moves down, and another layer of powder is spread over the printed surface. This process is repeated until the entire part is finished. Some printed parts, such as sand-casting cores and molds, can be used directly after the binder jetting process. However, most applications require a postprocessing step because binder-jetted parts typically have poor mechanical properties and high porosity upon exiting the printer.

In postprocessing, metallic parts are often sintered at high temperatures to improve the adhesion between metal particles or infiltrated with a low-melting-point metal such as bronze [12]. Similarly, ceramic parts usually undergo sintering and infiltration to enhance their mechanical strength and reduce porosity from binder removal. Additionally, applying an acrylic coating to the printed parts is common practice to improve visualization and enhance the vibrancy of colors in multicolor printed objects. The photos in Fig. 1.8B illustrate the major steps in the binder jetting production process: (i) 3D printing, (ii) part removal from the printer, (iii) binder removal and sintering, and (iv) the final product after cleaning and postprocessing.

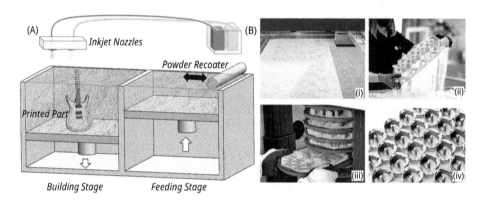

Fig. 1.8: (A) Schematic of binder jetting technology and (B) step-by-step photos of the binder jet printed part: printing, part removal, sintering, and finished product. Photos were reprinted with permission from Desktop Metal, Inc.

Binder jetting is a great choice for applications that require appealing aesthetic properties and parts for visualization, such as architectural models, toys, and figurines. This technique can produce full-color 3D-printed parts at high resolution, similar to material jetting. Key advantages of binder jetting include the low cost of powder feedstock and the high speed of the printing process.

While binder jetting is generally not suitable for functional applications due to the brittleness of the parts, metal-based binder jetting components can achieve relatively good mechanical properties if proper postprocessing steps such as infiltration or sintering are employed. Additionally, since the printing occurs at room temperature, dimensional distortions and warping caused by thermal stresses are not an issue, allowing for larger build volumes compared to other AM technologies. This capability enables the production of multiple parts and large objects, such as casting molds. Similar to powder bed fusion systems, binder jetting does not require support structures, as the surrounding powder provides the necessary support during the printing process.

1.3.4 Material extrusion

According to the ISO/ASTM definition, "material extrusion is an additive manufacturing process in which material is selectively dispensed through a nozzle or orifice" [13]. Material extrusion is the most widely adopted AM technology due to its simplicity, broad range of material options, low costs for both printers and feedstock materials, and the functional capabilities of the printed parts. Although there are several different extrusion processes, they can be primarily categorized into two major groups: fused filament fabrication (FFF) and paste extrusion.

1.3.4.1 Fused filament fabrication (FFF)
In this technique, a filament preform made of thermoplastic material is fed into an orifice, where it is melted and extruded through a nozzle. Once deposited, the material cools rapidly and solidifies into a single line (also known as a "road"), as illustrated in Fig. 1.9A.

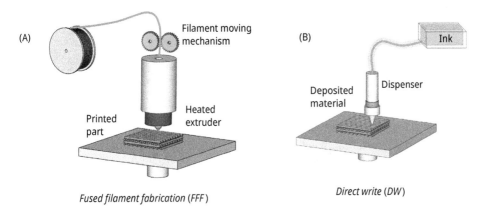

Fig. 1.9: Schematic of extrusion technology: (A) fused filament fabrication (FFF) and (B) direct write (DW).

The nozzle and/or the printing platform can move along the x–y- and z-axes simultaneously to deposit the material in a 3D configuration that matches the digitally designed CAD model. Most desktop AM machines in use today employ this method, which is why it is often the process people refer to when they mention 3D printing due to its widespread adoption.

1.3.4.2 Paste extrusion
This method is commonly referred to as direct write (DW), liquid deposition modeling, or robocasting. In this extrusion technique, a viscous, paste-like material is extruded through a nozzle (or tip). Unlike melting-based extrusion (FFF), the material in

DW is a viscous fluid rather than a solid when deposited on the print bed. This method relies on the fluid's yield stress to create self-supporting structures. The fluid's viscosity and yield strength can be adjusted using rheological modifiers such as nano-clay [14] and fumed silica [15]. The extrusion can be carried out using either a pressure-controlled or displacement-controlled system, as shown in Fig. 1.9B. In a displacement-controlled extrusion, a stepper motor moves the piston plunger downward toward the print bed to extrude the material. In pressure-controlled systems, a pump applies pressure directly to the material to push it toward the nozzle. The pressure level or the microstep increments of the stepper motor can be adjusted to control the extrusion speed.

In DW, the rheological properties of the extruded material play a crucial role in determining printability. Highly viscous pastes are preferred as printing materials because they can resist deformation after printing and maintain their shape. Shear thinning is a commonly observed behavior in DW, where material viscosity decreases as a function of shear rate, as illustrated in Fig. 1.10A. In shear-thinning materials, the viscosity drops significantly during extrusion, allowing for high flowability as the material is shaped into the desired geometry. After extrusion, the material's viscosity ideally recovers quickly, enabling it to retain its shape under gravitational forces without sagging. During the extrusion process, the shear rate ($\dot{\gamma}$) is maximal at the walls of the extruder and can be estimated as follows:

$$\dot{\gamma}_{max} = \frac{4\dot{Q}}{\pi r^3} \tag{1.1}$$

where r is the nozzle radius and $\dot{Q}$ is the volumetric flow rate, calculated as $\dot{Q} = Vr^2$, with V being the printing speed. Typical values of the shear rate during DW process is $50–100 \text{ s}^{-1}$. In Newtonian fluids, however, viscosity is constant and does not vary as a function of the shear rate.

Printability and the rheology relationship can be better identified by quantifying the storage and loss moduli of the DW extruded materials. The storage modulus (G') relates to the material's ability to store elastic energy and the loss modulus (G'') is the material's ability to dissipate stress through heat. Figure 1.10B is a representative plot for a viscous, shear thinning paste material, indicating the variation of storage and loss moduli as a function of shear stress. In Fig. 1.10B, the storage modulus is higher than the loss modulus under low shear, indicating that the material exhibits solid-like elastic behavior. On the other hand, under high shear, the loss modulus can exceed the storage modulus, signifying a liquid-like behavior of the printed material. At the crossover (or gelation) point, where the storage and loss moduli are equal, the material's yield stress can be measured. A high yield stress is desirable for extruded materials because it allows them to resist deformation and maintain their shape without sagging. However, materials with high yield stress require greater extrusion pressures or forces to be pushed through the nozzle orifices.

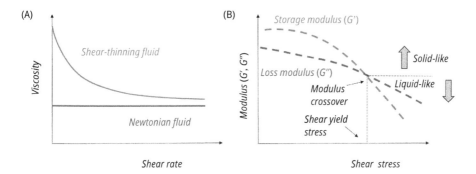

Fig. 1.10: Rheology considerations in direct write: (A) comparison of a Newtonian fluid and shear thinning fluid and (B) typical variation of loss and storage moduli as a function of shear rate during extrusion.

In the DW method, the entire printed structure is deposited at room temperature, significantly reducing the influence of mechanical properties on thermal printing history and spatial heating paths [16]. A wide range of materials can be additively fabricated using this method, including thermosetting polymers, thermoset composites, ceramics, ceramic composites, and conductive metal inks such as copper and silver. Postprocessing is typically required to enhance the mechanical strength of the printed components. For thermoset materials, curing is done at 100–200 °C, while ceramic-based pastes require heat treatment at elevated temperatures of 1,000–1,500 °C to facilitate the sintering of ceramic particles. Postprocessing is often unnecessary for printed conductive inks, as they are typically used in planar geometries where high strength is not critical [17].

DW expands the applicability of AM to liquid-based materials, composites, and ceramics. Compared to filament-based FFF techniques, DW is relatively new and still under development. It is currently limited to smaller volumes and height-constrained structures, as taller 3D structures require feedstock with extremely high viscosity and yield strength, or the use of support structures.

1.3.5 Powder bed fusion

Powder bed fusion is an AM technique in which a heat source is used to melt and fuse powder particles together to create a 3D object. The heat source used in powder bed fusion can be a laser, an electron beam, or a heat lamp. As illustrated in Fig. 1.11, a thin layer of powder is spread across the build platform, and heat is applied to selected areas to fuse the powder together in those regions. Once the layer is fused, the platform moves downward, and another layer of powder is spread over the previous layer using a roller. The heat is applied again to fuse the new layer. This process is repeated layer by layer until the entire model is constructed. After fabrication is complete, the unfused powder is removed, and the printed part is separated from the build plate.

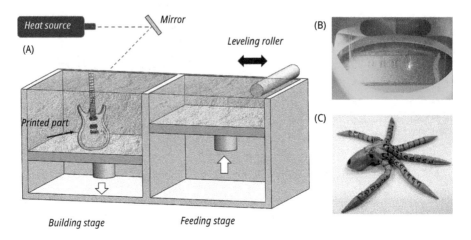

Fig. 1.11: (A) Schematic of powder bed fusion technology, (B) laser sintering in progress, and (C) octopus figurine fabricated via powder bed fusion (SLS).

There are multiple limitations of powder bed fusion technologies. First, high temperatures and heat introduced into the part may cause warpage and residual thermal stresses. In addition, powder bed fusion is one of the slowest AM techniques since it commonly includes powder preheating (to speed up the process/enhance powder fuse), vacuum generation, and material cooling-off period. Postprocessing is also common, adding the manufacturing time and cost. Since the parts are made fusing material powder together, surface quality depends on the grain size of the powder and would be very similar to manufacturing processes like sand or die casting [18]. The parts are manufactured over a build plate; hence, support removal postprocessing is necessary. Since material melting is necessary (partially or fully), this technique uses a significant amount of energy to create parts compared to other AM techniques.

Powder bed fusion is particularly effective at producing overhangs and downward-facing surfaces without the need for additional support structures, as the unfused powder serves as a natural support. This technology enables the fabrication of ceramics, polymers, and metals in complex 3D geometries. In recent years, the cost of powder feedstock and powder-based fusion machines has decreased significantly. Additionally, unfused powder can be recycled, provided contamination and degradation are carefully monitored to maintain the quality of parts. Selective laser sintering (SLS) and selective laser melting (SLM) are the most commonly used powder bed fusion technologies. In SLS, powders are heated near their melting point, leading to partial melting and sintering, while in SLM, the material is heated above its melting point, resulting in a complete melt and the formation of a homogeneous, nonporous structure.

Despite the numerous benefits mentioned above, powder bed fusion technology does have some limitations. High temperatures and heat can cause warping and residual thermal stresses. The process is also one of the slowest AM techniques due to the need for powder preheating, vacuum generation, laser raster scanning, and cooling periods. Postprocessing is often required, adding to the overall manufacturing time and cost. The surface quality of parts depends on the grain size of the powder and is comparable to other manufacturing processes like sand or die casting. Support removal postprocessing is necessary since parts are built on a plate, and the technique also consumes a significant amount of energy due to the need for material melting.

1.3.6 Directed energy deposition

Directed energy deposition (DED) is an AM process in which metal wire or powder is melted and deposited onto a build plate or an existing part using an energy source, as shown schematically in Fig. 1.12A. A typical DED system consists of a nozzle mounted on a multiaxis robotic arm within a closed frame. This nozzle deposits the molten material onto the workpiece surface, where it solidifies. The multidirectional movement of these robotic arms allows objects to be built quickly from various angles, provided the area is within the arm's reach. The process is similar in principle to material extrusion techniques, but unlike FFF, where filament is melted before deposition, in DED, the material is melted directly at the deposition surface. Moreover, DED systems can operate with up to five motion axes, compared to the three typically used in most FFF machines. Material deposition in DED occurs at high speed, making it one of the fastest AM technologies available [19]. DED can produce fully dense parts with complex geometries without requiring support for overhanging features. Additionally, it is particularly effective for adding metal to the existing parts, making it a preferred technique for welding and repair applications.

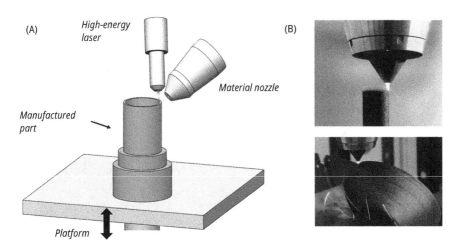

Fig. 1.12: (A) Schematic of DED additive manufacturing process and (B) metallic parts fabricated via DED technology. Published with permission from AddUp/BeaAM Inc.

Various metals such as aluminum, copper, titanium, tantalum, copper-nickel alloys, and steel alloys can be 3D printed using the DED technique. However, a primary drawback of this process is the rough surface finish caused by the high printing speed and the large metal melt pool size in this AM technology. As the melt pools cool, they create uneven surfaces, and therefore, most DED-produced parts require postprocessing, often in the form of secondary machining, to achieve a smoother finish. Additionally, the DED process involves localized heating at elevated temperatures, which can lead to the formation of thermal stresses in the printed object. To mitigate these stresses and other heat-related issues, postprocessing techniques such as hot isostatic pressing and heat treatment are commonly applied. Figure 1.12B illustrates the DED process applied to various metallic parts by AddUp/BeAM Inc.

1.3.7 Sheet lamination

Sheet lamination is a process that stands out from other AM techniques because the feedstock is not a liquid resin, filament, or powder. Instead, as the name suggests, it involves bonding layers of material sheets to build a 3D object. This method can be applied to a wide range of materials, including paper, polyvinyl chloride (PVC) polymers, metals, and ceramics. As illustrated in Fig. 1.13A, the process begins by laminating or bonding the sheets together, followed by cutting the 2D outline of the desired part using a laser or blade. In some cases, traditional CNC milling can be used to machine away the material. After cutting or machining, the next sheet is placed on top of the previous layer, and the process of bonding and cutting is repeated until the complete 3D object is formed.

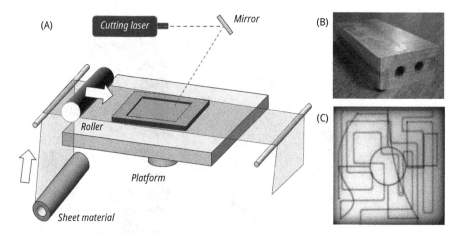

Fig. 1.13: (A) Schematic of sheet lamination technology and (B) parts fabricated via sheet lamination technology. Images were reprinted with permission from [20].

When paper sheets are used as a feedstock in sheet lamination, bonding is typically achieved with an adhesive or glue. For PVC sheets, thermoplastic polymers are melted to bond the layers together. In metal lamination, a localized energy source such as laser or ultrasonic waves is employed to fuse precision-cut metal sheets into a 3D object. Ultrasonic bonding, also known as ultrasonic AM or ultrasonic consolidation, is the most common method for metal sheet lamination. This process bonds metal sheets at room temperature through the application of ultrasonic waves and mechanical pressure, creating diffusion at the atomic level rather than relying on melting.

Sheet lamination is ideal for producing low-cost, full-color prints without high geometric complexity. The method also allows for the creation of parts with internal structures without the need for support. It is particularly beneficial for metal fabrication processes, where the thermal stress associated with melting, such as powder bed fusion techniques, would be problematic. While thermal stress is minimal in sheet lamination, an external cooling procedure is often applied between the lamination of each layer to further mitigate any residual stress from the bonding process.

Compared to other AM technologies, sheet lamination is less commonly used due to the high cost of the systems, specialized applications, and limited geometric freedom of the parts produced. Figure 1.13B showcases an aluminum heat exchanger block featuring internal channels that can be fabricated with varying complexity. This X-ray image of the sheet-laminated component demonstrates the capability to create intricate internal flow paths, which are challenging or impossible to achieve with traditional manufacturing methods.

As previously mentioned, each of the seven AM technologies is distinct, utilizing different methods to fabricate 3D objects layer by layer. Each technique comes with its own set of benefits and limitations concerning applicable materials, costs, resolution, speed, and build volume. Table 1.1 summarizes and compares the major properties of each of these seven AM techniques.

Tab. 1.1: Comparison of AM technologies.

AM method	Materials	Resolution	Advantages	Disadvantages
Vat polymerization	UV-curable photopolymers (acrylates/epoxides)	25–100 μm	Excellent surface quality, high resolution, no porosity, and isotropic properties	Limited mechanical properties and aging
Material jetting	UV-curable photopolymers (acrylates/epoxides)	25–100 μm	Fast, allows multimaterial, multicolor printing	Low viscosity ink required
Binder jetting	Starch PLA Metals Ceramics	50–100 μm	Fast, allows multimaterial printing	Limited mechanical properties and rough surfaces

Tab. 1.1 (continued)

AM method	Materials	Resolution	Advantages	Disadvantages
Powder bed fusion	Thermoplastics (PA6, PA12, and PEEK) Metals (stainless steel and titanium)	50–100 μm	Best mechanical properties, less anisotropy, and applicable to broad range of materials	Rough surfaces, thermal stress in the printed part, and poor powder reusability
FFF	Thermoplastic polymers (ABS, PLA, nylon, PC, PETG, and PEEK)	100–150 μm	Compact, inexpensive 3D printers, and good resolution	Limited materials, high temperature, porosity, and anisotropy
Direct write	Thermosets (epoxy, cyanate ester, and bismaleimide) Composites Hydrogels Biomaterials	100 μm to 1 cm	Broad range of materials	Low surface quality and room temperature printing
Sheet lamination	PVC Paper Sheet metals	200–300 μm	Low cost and low thermal stress	Limited geometrical freedom
Directed energy deposition	Metals (aluminum, copper, titanium, tantalum, copper, nickel, and steel alloys)	200–300 μm	Fastest AM technology	Poor surface finish, requires postprocessing, and thermal stresses in the parts

1.4 Timeline/history of additive manufacturing

To forecast the future of AM technology, it is crucial to reflect on its history and the transformative impact it has had on our lives. Originating 40 years ago, this technology has already revolutionized various sectors, including manufacturing, biomedicine, architecture, automotive, and aerospace. In short, its influence is widespread. Figure 1.14 outlines significant milestones in AM since the first patent was filed in 1984. A brief description of these key events is provided below:

– The idea of AM was conceived in the 1970s with the development of computers, CAD systems, laser technology, and micron-resolution stepper motors. However, the timeline of AM really starts with the first patent filed by Charles W. Hull in 1984 on SLA process. This technology was then commercialized when he founded the company 3D Systems in 1986.

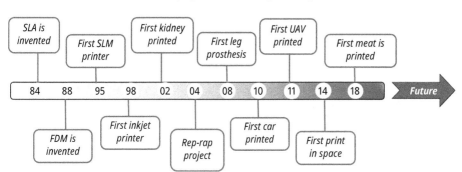

Fig. 1.14: Timeline of additive manufacturing technology development.

– In 1988, 3D Systems developed the STL file format for CAD models for slicing of the 3D models into 2D layers. Since then, the STL file format has been considered as the main file format for layer-by-layer manufacturing and greatly assisted for the merge of AM technologies.

– In 1988, Scott Crump invented the FFF technology using wax and a hot glue gun. This extrusion technique was named as fused deposition modeling and patented by the company Stratasys founded by Scott Crump, which later became one of the largest AM companies in the world.

– In 1989, the first patent for powder bed fusion technologies, including SLS, SLM, and electron beam melting, was granted based on the research conducted at the University of Texas at Austin. Although the patent was awarded in 1989, SLM and direct metal laser sintering were later developed in Germany in 1995 through collaboration with the Fraunhofer Institute, EOS, and other partners [21]. These technologies dramatically transformed AM particularly for metallic materials.

– The first commercial binder jetting machines were released after 1994, with the multijet printing process introduced by 3D Systems in 1996. In 1998, Objet Geometries developed PolyJet technology [22]. These binder jetting innovations enabled the production of multicolor, highly realistic prints with high resolution, significantly impacting the AM industry.

– In 2002, AM was applied on biomedicine by 3D printing a miniature kidney model. This functional kidney was able to filter blood and produce urine in an animal model. After 17 years, in 2018, the 3D-printed kidney saved the life of a 2-year-old boy as surgeons used this 3D-printed kidney model to perform the organ transplant [23].

– Implementing AM on biomedical field expanded dramatically after 2000s. In 2003, Thomas Boland from Clemson University patented the use of inkjet printing for cells and cell constructs [24]. This process allowed the deposition of cells into organized 3D matrices placed on a substrate.

– In 2005, Adrian Bowyer at the University of Bath initiated the RepRap project to make the FFF technique accessible to all. He developed an open-source 3D printer capable of partially replicating itself (RepRap) [25]. This open-source design, combined with the expiration of patents, led to the emergence of new AM companies, which reduced machine costs and accelerated advancements in AM technologies.

– In 2008, the first prosthetic leg was successfully used by an individual. All components of the prosthetic – leg, knee, foot, and socket – were fabricated additively, eliminating the need for assembly.

– In 2010, the entire body of a car, including its glass panel prototypes, was fabricated with the AM processes. This groundbreaking vehicle, named Urbee, was a collaborative effort between the Winnipeg engineering group Kor Eco-Logic and Stratasys [26]. Additionally, in 2014, Oak Ridge National Laboratory (ORNL) designed and 3D printed the Shelby Cobra electric car using FFF technology, employing a large-area AM (BAAM) printer developed by ORNL and Cincinnati Inc.

– In 2011, engineers at the University of Southampton successfully additively fabricated the first unmanned aerial vehicle (UAV). The entire structure of this UAV, including its wings, control surfaces, and access hatches, was produced through AM.

– In 2014, history was made when the first object was printed in space. The International Space Station's newly installed 3D printer successfully manufactured this groundbreaking item, marking the first-time AM was conducted beyond the Earth. NASA's FFF 3D printer was developed in collaboration with the startup Made In Space, founded in 2010. The goal of the 3D printer was to explore the possibility of producing essential replacement parts on the station, thereby eliminating the high costs associated with shipping them via rocket.

– In 2018, an Italian bioengineer Giuseppe Scionti developed a technology to create fibrous, plant-based meat analogs using a custom AM system based on the DW method. The additively manufactured samples exhibited textures and nutritional values comparable to those of natural meat [27]. This innovation represented a significant advancement in the application of AM within the food industry and a major step toward addressing global hunger issues.

Reflecting on the timeline of AM and the remarkable achievements in the past 40 years is truly inspiring. This technology has revolutionized the manufacturing industry and significantly impacted our lives. So, what does the future hold? What can we expect from the evolution of AM technology? While predicting the future is challenging, the current trajectory suggests that it will continue to expand, further displacing traditional manufacturing methods and positively transforming our world.

2 Additive manufacturing of polymers

2.1 Classification of polymers

Polymers are a class of engineering materials that are composed of chains of repeating chemical units called monomers. In fact, the word "polymer" is derived from two Greek words, "poly" and "mer," which means "many" and "units," respectively [28]. The repeating units can range from simple structures with just a few atoms to more complex, ring-shaped configurations containing multiple molecules as shown in Fig. 2.1. The figure also shows the repeating units of some of the most common polymers, including polyethylene, polyvinyl chloride, polytetrafluoroethylene, and polystyrene (PS). Different atoms (e.g., H, F, and Cl) or molecules (e.g., CH_3, C_6H_5, or a benzene ring) can be part of the repeating unit. In addition to homopolymers, which contain a single type of repeat unit, some polymers have two or more distinct repeat units known as copolymers.

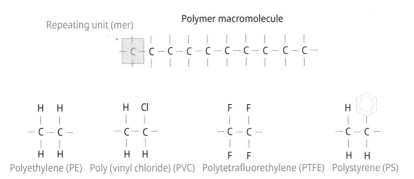

Fig. 2.1: Repeat units for common polymeric materials.

Natural polymers, such as proteins, cellulose, enzymes, starches, and nucleic acids, are found in living organisms and play essential biological roles. Other natural polymers, like wood, rubber, cotton, wool, leather, and silk, are derived from plants and animals. However, it was the advent of synthetic polymers that truly transformed our world over the past 50 years. Synthetic polymers, including plastics, rubbers, and epoxies, are made from petroleum-based organic molecules. These materials are cost-effective to produce, and their properties can be tailored to outperform natural polymers in many ways. In various applications, synthetic polymers have replaced traditional materials like metals and wood, offering advantages such as lower cost, lighter weight, and resistance to corrosion and chemicals.

Polymers can be classified based on various molecular characteristics, including repeat unit composition, molecular size (molecular weight), molecular shape (e.g., chain twisting and entanglement), and degree of crystallinity. However, in this book,

https://doi.org/10.1515/9781501520242-002

we will classify polymers by their molecular structure, as this classification primarily influences their physical properties and manufacturability. As illustrated in Fig. 2.2, polymers are divided into three main categories – thermoplastics, thermosets, and elastomers – based on their molecular structure. Each group exhibits unique physical properties due to its distinct microstructure. The next section will explore these categories in detail, along with the additive manufacturing (AM) technologies used to fabricate them.

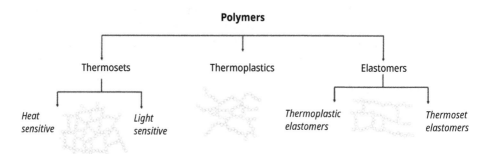

Fig. 2.2: Classification of polymeric materials.

2.1.1 Thermoplastics

Thermoplastic polymers have linear or slightly branched molecular structures, as shown in Fig. 2.2. The molecules in a thermoplastic polymer are held together by relatively weak intermolecular forces. Due to these weak interactions, thermoplastic materials soften (and eventually liquefy) when heated and return to their hardened state upon cooling. As temperature increases, intermolecular bonding forces weaken due to increased molecular motion, allowing adjacent chains to move more freely when stress is applied. This characteristic makes softened thermoplastics easy to reshape into their desired forms. Thermoplastics are widely used in applications such as toys, food packaging, thermal insulation, machine parts, and credit cards, as they can be repeatedly molded through heating.

2.1.2 Thermosets

Unlike thermoplastics, which are held together by weak intermolecular forces, thermoset polymers consist of monomer chains that are cross-linked to form a rigid, three-dimensional network. During the curing process, a thermoset resin (typically in liquid form) is combined with a catalyst or exposed to external energy, triggering a chemical reaction that permanently links the monomer chains. This cross-linking restricts the molecular chain movement, resulting in a durable material that does not soften or melt with heat, as thermoplastics do. Thermoset polymers are categorized into two main types: heat-sensitive and light-sensitive. Heat-sensitive thermosets use thermal energy to initiate cross-linking, while light-sensitive resins rely on light sources, such as ultraviolet (UV) or visible light, as described in Section 1.3.1. Due to their stability under heat, thermoset polymers are ideal for applications that require structural integrity at high temperatures. Compared to thermoplastics, thermosets are typically stronger, more heat-resistant, and often have a superior aesthetic finish. However, unlike thermoplastics, thermosets are not recyclable, as they cannot be melted and reshaped once cured. This property makes them less adaptable for recycling, but their durability and cost-effectiveness make them valuable for many industrial applications.

2.1.3 Elastomers

Elastomers are cross-linked polymers, but unlike thermoset polymers, they have a very low cross-link density (see Fig. 2.2). This structure allows the polymer chains some freedom of movement while preventing them from permanently shifting due to the cross-links between molecules. As a result, elastomers are highly stretchable and can quickly return to their original shape once the applied stress is removed. Elastomers are further classified into thermoplastic and thermoset elastomers. Thermoplastic elastomers, which melt when heated, are commonly used in thermal manufacturing processes such as injection molding, as they can be shaped using heat. Thermoplastic polyurethanes (TPUs), a key group of thermoplastic elastomers, are widely used in applications such as foam seating, seals, and gaskets. In contrast, thermosetting elastomers do not melt or soften with heat. Rubbers, the most commonly used thermosetting elastomers, are valued for their flexibility and durability. They are widely used in products such as tires, tubes, hoses, window profiles, gloves, balloons, conveyor belts, and adhesives.

2.2 Selection of polymers for additive manufacturing

The ability of melting and shaping polymers under heat is a key selection criterion for AM of polymers as described above. However, beyond melting capability and heat response, several other factors also impact material selection, as illustrated in Fig. 2.3:

– Printability: For effective AM process, the selected polymer must exhibit consistent flow during printing. It should also adhere well to the print bed and previously printed layers to ensure structural integrity and accurate geometry.
– Mechanical performance: Mechanical properties like strength, stiffness, elongation at break, and impact resistance are crucial for determining the functionality of the printed part. Polymers respond differently under various loading conditions; therefore, selecting the appropriate type and grade of polymer is essential for reliable performance.

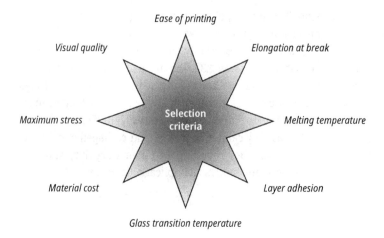

Ease of printing

Visual quality

Elongation at break

Maximum stress

Selection criteria

Melting temperature

Material cost

Layer adhesion

Glass transition temperature

Fig. 2.3: Physical properties of polymers affecting their selection process.

- Visual quality: The aesthetic appearance of polymers varies widely. While surface finish largely depends on the fabrication method and process settings, factors such as polymer type, color, and microstructure also influence the visual quality of the final product.
- Resistance to moisture absorption: Unlike metals and ceramics, polymers can absorb moisture in humid environments, which can cause defects during manufacturing. Some polymers, like nylon, are particularly moisture-sensitive and may require special storage. Moisture absorption can also lead to warping, hygromechanical stresses, and cracking in printed parts.
- Material toxicity: The toxicity of polymers used in AM is an increasing concern. Certain polymers, particularly thermoset resins, may contain toxic compounds that can emit harmful fumes during curing and potentially leach into liquids. Incomplete curing may affect biocompatibility, so nontoxic, biodegradable polymers are preferred for safety and environmental considerations. Polylactic acid (PLA), a biodegradable thermoplastic, is a commonly used polymer in AM.

2.3 Additive manufacturing of thermoplastic polymers

PLA, acrylonitrile butadiene styrene (ABS), polyethylene terephthalate glycol, high-density polyethylene, polycarbonate (PC), and nylon are among the most commonly used thermoplastics in AM. These materials have relatively low melting temperatures, making them suitable for fabrication through fused filament fabrication (FFF), as discussed in Chapter 1. For applications requiring higher temperature resistance, special high-melting thermoplastics such as polyetherimide (ULTEM), polyether ether ketone (PEEK), and polyetherketoneketone (PEKK) are preferred. These materials can reach the melting temperatures of up to 400 °C and also exhibit high glass transition temperatures. Glass transition temperature is critical for end-use applications, as it marks the threshold above which polymers transition from a glassy (or crystalline) state to a rubbery state. The maximum operating temperatures for polymers are typically kept below their glass transition temperatures. The high-performance polymers mentioned above require specialized printing systems capable of melting and extruding them onto the print bed at elevated temperatures.

In addition to FFF, thermoplastics can also be processed through powder bed fusion methods. Nylon, for instance, is the dominant material used in selective laser sintering (SLS), accounting for nearly 90% of SLS manufacturing [29]. Other materials, such as PSs and polyaryletherketones, are used in smaller amounts. Unlike FFF, SLS-compatible thermoplastics are limited due to specific material requirements. These include precise particle shape, powder distribution, and thermal, rheological, and optical characteristics, which collectively determine the successful SLS implementation [30]. Consequently, only a few thermoplastics are suitable for SLS to date. Table 2.1 lists the glass transition and melting temperatures of common thermoplastics used in FFF and SLS AM processes.

Ultimate tensile strength and elastic modulus are key properties that define the mechanical performance of polymers. Figure 2.4 summarizes these properties for additively manufactured thermoplastic polymers. As shown, there is a significant variation in the reported results, highlighting the influence of printing parameters, material composition, and environmental factors on the mechanical properties. High-temperature

Tab. 2.1: Cost, maximum service temperature and coefficient of thermal expansion (CTE) of common thermoplastic materials for additive manufacturing.

Polymer type	AM method	Extruder temperature	Maximum service temperature	Cost ($/kg)	CTE (μm/m °C)	References
PLA	FFF	190–220	52	10–40	68	[31]
ABS	FFF	220–250	98	10–40	90	[31]
PC	FFF	260–310	121	40–75	69	[31]
PET	FFF	230–250	73	20–60	60	[31]
PVA	FFF	185–200	75	40–110	85	[31]
ULTEM	FFF	330–350	174.4	150–170	52.7	[32]
PEEK	FFF	390–400	143	150–600	54–70	[32–34]
PEKK	FFF	375–390	161	200–400	23–27	[32, 35]
Nylon	FFF/SLS	220–270	80–95	25–65	80–100	[32, 36]
Polyimide	SLA/DW	–	360–410	>500	35.9	[37–39]
Epoxy	DW	–	166–206	50–100	45–65	[36]
Cyanate ester	DW	–	280	>65	30–45	[36, 40]

PLA, polylactic acid; ABS, acrylonitrile butadiene styrene; PC, polycarbonate; PET, polyethylene terephthalate; PVA, polyvinyl alcohol; ULTEM, polyetherimide; PEEK, polyether ether ketone; PEKK, polyetherketoneketone; FFF, fused filament fabrication; SLS, selective laser sintering; SLA, stereolithography; DW, direct writing.

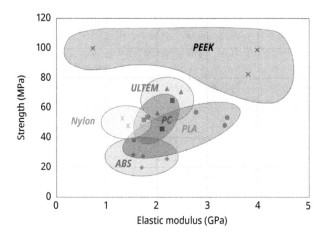

Fig. 2.4: Tensile mechanical properties of additively fabricated thermoplastic polymers used in AM.

thermoplastic materials, such as ULTEM and PEEK, offer superior strength compared to other additively manufactured thermoplastic materials. However, their higher material cost and the requirement for specialized printing systems capable of melting them at elevated temperatures are notable drawbacks of these materials.

In addition to tensile strength and elastic modulus, elongation at break and fracture toughness are crucial mechanical properties for specific applications. Polyamides, such as nylon, and polycarbonates are known to exhibit high elongation at break, exceeding 25% [41]. Despite its excellent fracture resistance, ease of fabrication, and low cost, nylon is prone to moisture absorption, requiring careful storage of the material feedstock in dry conditions before manufacturing. Using moist nylon filament or powder feedstock can result in defects in the final parts. Along with nylon and polycarbonate, ABS has also been reported to exhibit high elongation (~20%) at fracture in certain studies [42, 43]. However, ABS typically has lower tensile strength compared to other commonly used thermoplastic polymers.

2.4 Additive manufacturing of thermosets

Thermosets are highly cross-linked polymers. Cross-linking can be achieved by applying UV light on photopolymer resins or by heating thermally curing resins. Photopolymers are typically processed using vat polymerization or material jetting techniques, where UV light is employed to cure the material. Thermally cured thermosets, on the other hand, are usually printed using direct write extrusion processes.

2.4.1 Additive manufacturing of photosensitive thermosets

Typical photopolymer materials used in vat polymerization consist of a mix of monomers, oligomers, photoinitiators, epoxies, and various additives, such as inhibitors,
dyes, and toughening agents, which help adjust the printability and physical properties of the photopolymers [6]. However, photosensitive resins have two main drawbacks that limit their widespread use as production materials: sensitivity to water
and humidity, and a tendency to age over time. Studies have shown that the mechanical properties of parts produced via vat polymerization change over time due to environmental factors such as temperature, moisture, and UV exposure [44–46]. Similarly,
research on photocurable thermosets fabricated using the material jetting process indicates that these parts exhibit considerable variability in tensile and compressive
properties and show anisotropy [47]. Similar to vat polymerization, material jetting-
produced parts also experience time-dependent changes in mechanical properties
due to aging [48], highlighting that aging is a result of changes in the polymer's molecular structure and cross-linking, rather than the fabrication method itself.

Considering the complexity of photopolymer chemistry and the variety of resins
offered by different manufacturers, making a generic comparison of mechanical performance is challenging. However, tensile mechanical properties for selected photosensitive thermosets are provided in Fig. 2.5, which demonstrates the range of properties reported in previous studies. This figure also reveals that the mechanical
properties of additively manufactured parts depend more on the chemical structure
of the polymer than on the manufacturing method. For instance, the same AM technology, such as material jetting, can be used to fabricate both soft (Durus, Stratasys)
and stiff (High Temperature, Stratasys) photopolymers using the same printer as the
figure indicates.

2.4.2 Additive manufacturing of heat-sensitive thermosets

Heat-sensitive thermosets can be additively manufactured through direct write extrusion without the need for UV light exposure, provided the material's viscosity is suffi-
ciently high. To adjust the viscosity, liquid thermoset resins are often mixed with rheology modifiers such as clay or silica nanoparticles. Additionally, UV light can be used
to enhance the solidification process during direct write printing [49].

In this case, photosensitive compounds must be added into the thermoset resin.
Direct write AM of different types of thermoset resins such as epoxies [14, 16], cyanate
esters [40], and polyimides (Kapton) [50] have been reported in the literature. Tensile
mechanical properties of the fabricated parts from some of these studies are shown
in Fig. 2.5.

Thermal curing of polymers is a widely used 3D printing technique to achieve
high mechanical performance and thermal stability. However, the curing process in

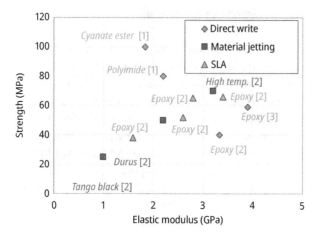

Fig. 2.5: Tensile mechanical properties of additively fabricated thermoset polymers.

heat-sensitive thermosets is significantly slower compared to light-curing photopolymers. To address this limitation, researchers have developed hybrid thermoset 3D printing methods that combine both curing mechanisms [51–53]. By mixing photopolymers with heat-sensitive thermosets, this approach enables faster curing while maintaining high mechanical performance. Cross-linking of the photopolymer ensures structural stability during the printing process, and subsequent thermal curing enhances the mechanical strength of the printed part.

More recently, a novel in situ curing approach, known as reactive curing or frontal polymerization, has been introduced for direct ink writing processes. Frontal polymerization involves initiating a polymerization front within the thermoset material through an applied stimulus. This front propagates due to the highly exothermic reaction chemistry, provided the exothermic heat generated is sufficient to sustain the reaction [54]. Typically, these polymerization fronts are triggered thermally, using heat, light, or radiation.

Figure 2.6 illustrates the integration of frontal polymerization into the direct ink writing process [55]. This method allows for rapid curing during printing, enabling freeform fabrication without the need for support structures. Another significant advantage of frontal polymerization is that it eliminates the need for postcuring in an oven, as complete curing occurs during the printing process. The resulting printed parts exhibit excellent mechanical stability, while the process itself minimizes energy consumption and manufacturing time, allowing for rapid and efficient production.

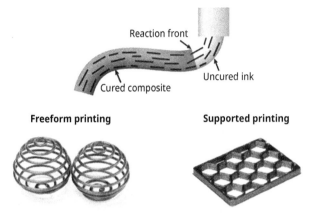

Fig. 2.6: Frontal polymerization process during direct write printing. Freeform printing can be achieved without any supporting structure. Figure was reprinted with permission from [55].

2.5 Additive manufacturing of elastomers

Elastomers are weakly cross-linked polymers that can be stretched extensively under mechanical loads. The AM of elastomers has seen growing applications across various industries due to their unique properties, including deformability, fracture resilience, and electrical and thermal insulation capabilities. As shown in Fig. 2.2, elastomers are classified into two categories: thermoset elastomers and thermoplastic elastomers. Thermoplastic elastomers behave similarly to thermoplastic polymers, as they soften and melt when heated. On the other hand, thermoset elastomers act like thermosets, as they can be cured or solidified by applying heat or light, irreversibly cross-linking the polymer chains. Photocurable elastomers can therefore be manufactured using vat polymerization (stereolithography, digital light processing, and continuous liquid interface production) and material jetting processes, much like photosensitive thermosets.

Several commercially available thermoset elastomer resins, including PDMS [56], Carbon EPU40, Stratasys Tango Plus, Formlabs Flexible, and Spot-A Elastic [57], have been developed. Despite the high-resolution printing capabilities of UV-curable resins and vat polymerization technology, the maximum elongation at break for these materials is currently limited to 170–220% in these AM techniques [57]. As an alternative to vat polymerization and material jetting, direct write techniques are also used to print various types of elastomers without the need for UV curing. Direct write AM can be applied to highly stretchable elastomers, such as EcoFlex, which has been reported to achieve up to 900% elongation at break [58]. While this technique offers greater flexibility in terms of the range of elastomeric materials that can be used, it sacrifices part complexity and geometric resolution compared to photopolymerization methods, as discussed in Chapter 1. Figure 2.7 provides examples of flexible elastomers fabricated by Formlabs Inc. via vat polymerization, showcasing their high flexibility and dimensional accuracy.

Fig. 2.7: 3D-printed structures made with Formlabs elastic resin using vat polymerization technology. Figures were reprinted with permission from Formlabs Inc.

The second class of elastomers, thermoplastic elastomers, are also widely used in AM applications. These materials can easily be melted in the print head and processed using the FFF method. TPU is the most commonly used elastomer for this technique, with reports of up to 700% elongation at break [59]. In addition to their high stretchability, elastomers produced through this process also exhibit excellent dimensional accuracy, typically ranging from 0.1 to 0.2 mm, as determined by the FFF technology.

3 Additive manufacturing of polymer composites

For rapid prototyping or applications where components experience low mechanical loads, neat thermoplastic and thermoset polymers are ideal material choices. However, when higher mechanical strength is needed for structural applications or when multifunctional properties (such as enhanced electrical/thermal conductivity, piezoresistivity, or energy storage capacity) are desired, polymer composites are preferred. To achieve these properties, the polymer matrix can be infused with powders, short fibers, or continuous fibers. Figure 3.1 classifies additively manufactured polymer composites based on the type of dopant or reinforcement material. Each category will be examined in the following section, with comparisons of additive manufacturing methods and physical properties for each composite group.

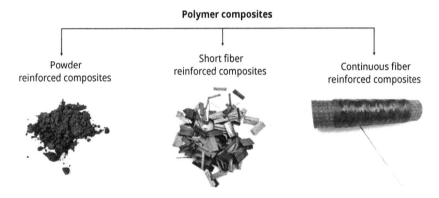

Fig. 3.1: Classification of additively manufactured polymer composite materials.

3.1 Additive manufacturing of powder-doped polymer composites

Infusing polymers with powdered materials is a common approach in composite manufacturing. These additives can be nanoscale, such as carbon nanotubes (CNTs), quantum dots, and graphene platelets, or microscale with low aspect ratios, such as carbon black (CB), metallic powders, or ceramic powders. Powder doping allows the tailoring of various properties in printed parts, including electrical conductivity, thermal conduction, shape memory, dielectric properties, piezoelectricity, and optical properties. Powder-doped composites can be produced using a variety of additive manufacturing methods – including extrusion (fused filament fabrication (FFF) and direct write), vat polymerization, binder jetting, material jetting, and powder bed fusion – thanks to the small size of the powder-based additives and their ease of integration with the polymer matrix.

https://doi.org/10.1515/9781501520242-003

One of the most beneficial applications of powder doping in polymer composites is in adjusting electrical properties. This has led researchers to explore various dopant effects on the conductivity of additively manufactured polymer structures. CNTs, for instance, have high electrical conductivities similar to metals. Chizari et al. [60] recently created highly conductive CNT/PLA (polylactic acid) nanocomposites for fabricating conductive scaffold structures through additive manufacturing, dispersing multiwalled CNTs (MWCNTs) at high concentrations (up to 40 wt%) in PLA using a ball milling method. Postiglione et al. [61] also employed PLA/MWCNT nanocomposites for additive manufacturing of conductive 3D structures via direct write methods, achieving a percolation threshold concentration of 0.67% CNTs with a conductivity of 10 S/m and maximum conductivity of 100 S/m at 5 wt% MWCNT.

Graphene, a 2D nanomaterial, is gaining substantial interest in advanced manufacturing due to its low resistivity, high thermal and electrical conductivity, and optical transparency. Numerous studies have successfully used graphene to create conductive FFF filaments, achieving high conductivity values (up to 166 S/m) in graphene–PLA FFF filaments [62]. The high cost of graphene continues to be a major obstacle to its broader application in feedstock materials.

CB is another conductive additive widely used in additive fabrication of polymer composites. CB is produced from the incomplete combustion of heavy petroleum products such as coal tar; therefore, it is readily available and inexpensive [63]. Low cost, chemical stability, and high conductivity make CB one of the most popular conductive additives. Kwok et al. [62] developed conductive polypropylene (PP)-based thermoplastic composites for FFF-based 3D printing of electrical circuits, achieving high conductivity (~200 S/m) with composites containing ≥30% CB by weight. This study also showed that additively manufactured composites containing over 25% CB by weight were suitable for fabrication and repair of the practical size electrical circuits.

In addition to electrical conductivity, thermal conductivity in polymer composites can also be significantly enhanced with conductive micro- and nanomaterials, as heat transfer within the material occurs through electrons, which are also the carriers of electric current:

$$\kappa = \kappa_e + \kappa_L \tag{3.1}$$

where κ is total thermal conductivity, κ_e is the electronic contribution, and κ_L is the lattice contribution on heat conduction within a material. As electrical conductivity of a material increases, electronic contribution in eq. (3.1) increases; therefore, thermal conductivity also increases. Recent studies indicate that conductive dopants in powder form, such as CNTs, graphene, CB, copper, bronze, magnetic iron, and stainless steel, can significantly improve the thermal conductivity of polymer composites [64, 65]. Consequently, polymer composites doped with conductive materials can transport heat more efficiently than undoped polymers, promoting a uniform temperature distribution across print layers. This improved heat distribution helps minimize thermal stresses, warpage, and spatial inconsistencies caused by high temperature gradients in components fabricated with these high thermal conductivity materials.

Piezoelectric materials can convert mechanical stresses, such as compression and tension, into electrical charges, and vice versa. This property enables their use in a variety of applications, from speakers and acoustic imaging to energy harvesting and electrical actuators. Traditionally, piezoelectric materials are manufactured by machining or mechanical dicing, which restricts the complexity and size of the elements due to the brittle nature of these materials. To overcome these limitations, Kim et al. [66] developed a novel approach to fabricate 3D piezoelectric materials using piezoelectric nanoparticles embedded in a photocurable PEGDA (polyethylene glycol diacrylate) polymer. In their study, barium titanate (BTO) nanoparticles were chemically modified with acrylate groups, which formed covalent bonds with the polymer matrix under light exposure. Using this method, complex 3D geometries (e.g., mushroom, cross, tapered cantilever, and microtubule shapes) were achieved with a piezoelectric coefficient of ~40 pC/N at a 10% BTO nanoparticle loading, which outperformed samples using unmodified BTO and CNTs.

Nanomaterials also play a role in tuning the optical properties of additively manufactured polymers. Due to their nanoscale size, zero-dimensional nanoparticles can absorb light at various frequencies based on their environment, making them popular for vibrant coloring in industries such as cosmetics, food, and textiles. Recently, carbon quantum dots (CQDs) of 2–3 nm in size were incorporated into photocurable resins to modify the optical response in additive manufacturing [67]. Using stereolithography (SLA), CQD-enhanced resins were shaped into the "Statue of Liberty" models, and, as shown in Fig. 3.2, the optical response to UV light of doped versus undoped (neat) samples revealed distinct differences, showcasing the potential for customizable optical properties.

As will be discussed in more detail in the next section, powdered doping materials typically lack the high aspect ratios needed for mechanical strength enhancement, or they do not provide reinforcing capabilities due to their nanoscale dimensions (e.g., CNTs). Consequently, to improve the strength and stiffness (modulus) of polymer composites, short and continuous fiber reinforcements on a larger scale are often preferred over nano/microscale doping materials.

While direct enhancement of tensile strength and stiffness through micro/nanoscale reinforcements is not feasible, CNTs can indirectly improve the mechanical strength of additively manufactured polymers. In a study by Sweeney et al. [68], CNT-coated thermoplastic filaments were shown to increase the strength of printed parts in the FFF process. This approach involves coating thermoplastic filaments with a thin CNT layer by dipping them into a MWCNT ink prior to 3D printing. These CNT-coated filaments are then extruded via FFF to form the final parts. Because CNTs are located on the filament's surface, they are deposited at the interfaces between printed roads during extrusion. When microwave heating is applied, CNTs within the polymer respond to the microwaves, causing localized melting at the in-

(a)   (b)

Fig. 3.2: Change of optical properties in additively manufactured photopolymer. (a) SLA printed Statue of Liberty of CQD-doped and -undoped (right) photopolymer (the scale bar is 5 cm) and (b) comparison of the specimen with a US dime. Figure was reprinted from [67].

terfaces between layers. This localized melting promotes material diffusion in these areas, strengthening layer adhesion. Since these regions are typically the weakest points in FFF-fabricated components, the local melting has been shown to increase both fracture and tensile strength in the 3D-printed materials. This process is known as locally induced RF welding.

3.2 Additive manufacturing of short fiber-doped composites

Additive manufacturing of short fiber-reinforced polymer composites (SFRPCs) is attracting increasing interest in composite fabrication due to its ability to create complex structural parts without requiring specialized manufacturing tools. SFRPCs can be produced using the same methods as those used for neat or powder-infused polymers. Beyond the ease of production, short fiber reinforcement can greatly enhance the mechanical properties of polymers, including tensile strength, elastic modulus, and fracture toughness, unlike powder-based reinforcements. The processing and additive manufacturing of short fiber-reinforced composites are highly dependent on the type of polymer matrix used – whether thermoplastic or thermoset – so each type will be explored separately.

3.2.1 Short fiber-reinforced thermoplastic composites

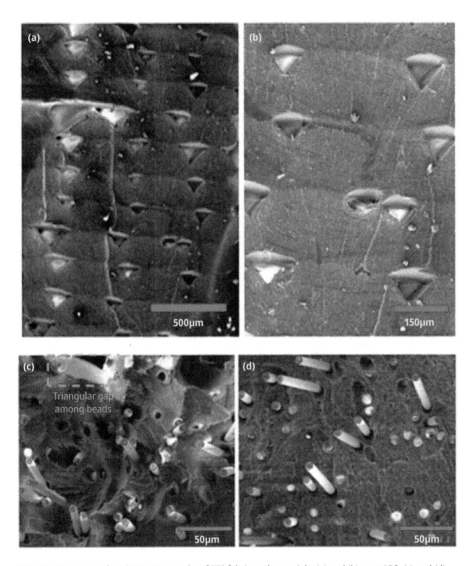

Fig. 3.3: Fracture surface SEM micrographs of FFF fabricated materials: (a) and (b) neat-ABS; (c) and (d) carbon fiber-reinforced composite. Figure was reprinted with permission.

Additive manufacturing of thermoplastic materials using FFF is a well-established technology, offering high-resolution and low-cost benefits, as discussed in Chapter 1. Multiple studies exist in the literature, where chopped polymer [69, 70], glass [71, 72], and carbon [73, 74] fibers were mixed with thermoplastic polymer resins such as PLA and ABS to prepare composite filament feedstock for FFF process. Short fiber-reinforced composites were then fabricated by melting these filaments and extruding them on the printbed using FFF. Carbon fiber is especially popular for reinforcement due to its high strength, chemical and thermal resistance, and low density. Previous studies on short carbon fiber-reinforced thermoplastics have shown substantial improvements in tensile strength and elastic modulus over unreinforced thermoplastic matrices. The enhanced stiffness in these reinforced composites is particularly advantageous, as it significantly reduces material distortion and warping during the 3D printing process [75].

The major reasons behind the low strength observed in short fiber-reinforced thermoplastic composites are the porosity between the printlines (unavoidable in FFF process) and poor fiber–matrix interfacial adhesion as evidenced by the protruding carbon fibers shown in Fig. 3.3.

While high fiber ratios (up to ~40%) are achievable in these composites, the maximum tensile strengths observed in FFF-printed short fiber composites are typically below 100 MPa, which is less than that of unreinforced thermoplastic PEEK (see Fig. 2.3). The primary reasons for this reduced strength are the inherent porosity between print lines in the FFF process and the limited fiber–matrix adhesion, evidenced by protruding carbon fibers, as shown in Fig. 3.3. A comparison between FFF-printed specimens and compression-molded samples by Tekinalp et al. [74] shows that while the tensile strength and modulus of additively manufactured composites with well-aligned fibers improve with increased fiber content, these properties still do not surpass those of compression-molded composites, where fibers are randomly oriented (Fig. 3.4).

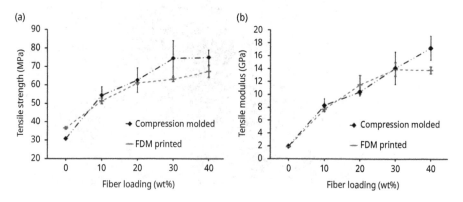

Fig. 3.4: Comparison of mechanical properties as a function of fiber loading in FFF versus compression molding: (a) tensile strength comparison and (b) tensile modulus comparison. Figure was reprinted with permission from [74].

Short fiber-reinforced thermoplastic filaments are now commercially available and can be used in any FFF 3D printing system, as long as the nozzle size and fiber volume fraction are appropriately chosen to prevent clogging of the extruder by short fibers during printing.

3.2.2 Short fiber-reinforced thermoset composites

As described above, weak adhesion between short fiber reinforcements and thermoplastic polymer is the main cause of low strength in these composites. To enhance adhesion between fiber reinforcement and the matrix, fibers are often coated with a few manometers thick surfactant layer, which is commonly known as sizing. This coating, usually polymer, chemically couples the matrix and the fiber creating a strong adhesion between these two components. Sizing chemistry is well-developed for coupling carbon fibers and thermoset resins, where liquid resin can wet the fiber surface and facilitate the chemical adhesion process. Additive manufacturing of liquid thermoset polymers can be achieved using direct write method as described in Chapter 1. Direct write additive manufacturing of carbon fiber-reinforced epoxy thermoset composites were introduced for the first time in 2014 by Compton et al. [16]. However, in this study, short fibers did not enhance the tensile strength of the fabricated composites, which was probably due to the weak adhesion between the thermoset matrix and the unsized fibers used in the composite. Recently, Pierson et al. [76] performed direct write manufacturing to fabricate carbon fiber/epoxy composites, where carbon fibers were sized for epoxy resin. In this study, tensile strength of 127 MPa was reached by using only 5.5% carbon fiber by volume as reinforcement. This was 236% increase in strength and 259% increase in elastic modulus compared to the unreinforced epoxy. In addition, comparison of the same material type fabricated via compression molding showed that additively manufactured samples showed 30% higher tensile strength and 47% higher elastic modulus. As shown by the SEM images in Fig. 3.5, the direct write process leads to the alignment of carbon fibers along the printing direction unlike compression molding where fibers are randomly aligned. In this study, the authors also observed that fiber aspect ratio, fiber volume fraction, and orientation significantly affect the mechanical properties (strength and modulus). The relationship between the fiber morphology and the mechanical properties is described in the next section in more detail.

As mentioned earlier, weak adhesion between the short fiber reinforcements and the thermoplastic polymer is a primary cause of low strength in these composites. To improve adhesion between the fiber reinforcement and the matrix, fibers are often coated with a thin layer of surfactant, known as sizing. This coating, typically made of a polymer, chemically bonds the matrix to the fiber, creating stronger adhesion between the two components. Sizing chemistry is well-established for coupling carbon fibers with thermoset resins, where liquid resin can wet the fiber surface and facilitate the chemical adhesion process. Additive manufacturing of liquid thermoset polymers can be achieved using the direct write method, as discussed in Chapter 1.

The direct write additive manufacturing of carbon fiber-reinforced epoxy thermoset composites was first introduced by Compton et al. in 2014 [16]. However, in their study, short fibers did not enhance the tensile strength of the fabricated composites, likely due to weak adhesion between the thermoset matrix and the unsized fibers. More recently, Pierson et al. [71] performed direct write manufacturing to fabricate carbon fiber/epoxy composites with sized fibers for epoxy resin. In their study, a tensile strength of 127 MPa was achieved with just 5.5% carbon fiber by volume, representing a 236% increase in strength and a 259% increase in elastic modulus compared to unreinforced epoxy. Furthermore, a comparison with compression-molded samples showed that the additively manufactured composites had 30% higher tensile strength and 47% higher elastic modulus.

As shown in the SEM images (Fig. 3.5), the direct write process results in the alignment of carbon fibers along the printing direction, unlike in compression molding where fibers are randomly oriented. The study also highlighted that fiber aspect ratio, fiber volume fraction, and orientation significantly impact the mechanical properties (strength and modulus), as further explored in the next section.

Fig. 3.5: SEM micrographs of the fractures surfaces: (A-C) random fiber alignment within compression molding sample and (D-F) uniformly aligned short carbon fibers in 3D-printed composite. Image was reprinted with permission from [76].

In addition to carbon fiber, short Kevlar (aramid) fibers are also employed as reinforcements in the additive fabrication of thermoset composites using the direct write method. A previous study [69] demonstrated that increasing the Kevlar fiber volume fraction enhances the composite's strength and modulus. Additionally, the study reported an increase in elongation at break due to the high flexibility of Kevlar fibers.

Consequently, additively manufactured Kevlar-reinforced composites show significant promise for applications requiring a combination of high strength, stiffness, fracture toughness, and impact resilience.

However, the maximum fiber loading in the thermoset composite studies mentioned above was limited to 5.5% by volume. Exceeding this threshold, particularly with carbon fiber, caused issues such as discontinuous material extrusion, nozzle clogging, and fabrication challenges. Developing innovative additive manufacturing techniques capable of processing short fiber-reinforced composites with higher fiber loadings could greatly expand the applicability and adoption of these materials across various industries.

3.3 Prediction of mechanical properties of short fiber-reinforced composites

For composite systems where short fibers are *perfectly aligned*, elastic modulus can be predicted by the well-known Halpin–Tsai analytical model [14, 77, 78] as follows:

$$E_L = \frac{(1 + 2s\eta_L f)E_m}{1 - \eta_L f} \quad \text{and} \quad \eta_L = \frac{(E_r/E_m - 1)}{(E_r/E_m + 2s)} \tag{3.2}$$

where s is the aspect ratio of the fibers, f is the fiber volume ratio, and E_r and E_m are the elastic moduli of the reinforcement and matrix, respectively. The elastic modulus (E_R) of a material with *randomly oriented* fibers can be determined using the moduli in the longitudinal and transverse directions, as follows:

$$E_R = \frac{3}{8}E_L + \frac{5}{8}E_T \tag{3.3}$$

where E_T is the elastic moduli in longitudinal and transverse directions and is obtained by

$$E_T = \frac{(1 + 2\eta_T f)E_m}{1 - \eta_T f} \quad \text{and} \quad \eta_T = \frac{(E_r/E_m - 1)}{(E_r/E_m + 2)} \tag{3.4}$$

According to these models, modulus in longitudinal direction increases by increasing the aspect ratio (s) of the fiber and the fiber volume ratio (f) within the composite. In the transverse direction, the fiber aspect ratio does not influence the composite modulus. Strength predictions can also be made using models described in [77, 78]. The ultimate strength of a material with aligned fibers is given as follows:

$$\sigma_L = \begin{cases} fs\frac{\sigma_m}{\sqrt{3}} + (1-f)\sigma_m, & s < s_c \\ f\sigma_r\left(1 - \frac{\sigma_r\sqrt{3}}{4s\sigma_m}\right) + (1-f)\sigma_m, & s \geq s_c \end{cases} \tag{3.5}$$

where s_c is the critical aspect ratio of the fibers, f is the fiber volume ratio, and σ_r and σ_m are the ultimate strengths of the fiber reinforcement and the matrix, respectively. In strength calculations, the critical aspect ratio (s_c) plays an important role. Critical aspect ratio (s_c) is defined as follows:

$$S_c = \frac{\sigma_r \sqrt{3}}{2\sigma_m} \tag{3.6}$$

The strength of a material, such as the one fabricated through compression molding with randomly oriented fibers, is calculated using the directional strengths:

$$\sigma_R = \frac{3}{8}\sigma_L + \frac{5}{8}\sigma_T \tag{3.7}$$

where σ_T is the strength in transverse direction and is obtained by

$$\sigma_T = \sigma_m \left(\frac{f}{\sqrt{3}} - f + 1 \right) \tag{3.8}$$

These analytical formulations suggest that if the fiber aspect ratio falls below a critical value (s_c), the fiber's strength does not significantly contribute to the composite's overall strength. This aligns with the understanding that increasing the aspect ratio enhances strength, explaining why tensile strength cannot be improved using powder reinforcements with low aspect ratios, as discussed in the previous section.

Experimental validation of these predictions was recently conducted by Pierson et al. [14], as illustrated in Fig. 3.6. The experimental results compared composites fabricated through additive manufacturing and compression molding techniques. For compression-molded composites with randomly oriented fibers, the test results closely matched the predictions for random fiber orientation, as shown in the figure.

However, the strength and modulus of additively manufactured composites were found to be lower than those predicted by the analytical models. This discrepancy may stem from the assumptions in models (3.2)–(3.8), which consider perfect fiber alignment within the composite. While SEM images showed a high degree of fiber alignment in the additively manufactured composites, perfect alignment was not achieved. Fiber misalignment and defects such as voids introduced during the printing process likely contributed to the reduced strength and stiffness of the additively manufactured thermoset composites.

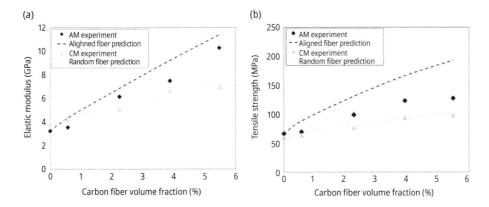

Fig. 3.6: Experimental and analytical model comparison of additively manufactured and compression-molded epoxy–carbon fiber thermoset composites for (a) elastic modulus and (b) ultimate tensile stress. Image was reprinted with permission from [76].

3.4 Alignment of short fibers within additively manufactured composites

The mechanical performance of SFRPCs is heavily influenced by fiber length (or aspect ratio) and the alignment of the short fibers within the composite structure. Recent theoretical [79, 80] and experimental [14] studies have shown that well-aligned short fibers significantly enhance mechanical performance in the alignment direction. Beyond mechanical properties, aligning fibers in specific orientations can also improve thermal properties and reduce moisture absorption in composites. Consequently, controlling fiber alignment has become a key area of research due to its substantial impact on the physical properties of additively manufactured materials. Various techniques have been explored to achieve fiber alignment in 3D printing technologies, such as applying electrical or magnetic fields or utilizing shear-induced fiber alignment. These methods are briefly described below:

- *Magnetic and electrical field alignment:* Magnetic and electric fields have been used to manipulate fiber alignment in liquid resins during processes such as stereolithography. For example, Nakamoto and Kojima [81] applied a magnetic field to align ferromagnetic γ-Fe_2O_3 fibers within a polymer. Similarly, conductive whiskers were used for alignment under an electric field, where the applied field created a moment that aligned the whiskers along the field direction in the liquid photopolymer [82]. However, these techniques require advanced mathematical modeling of the forces generated by external fields and are limited to materials with magnetic or conductive properties.
- *Shear-induced fiber alignment in direct write additive manufacturing:* In direct write processes, fibers suspended in a viscous polymer medium are forced through a con-

verging nozzle, where shear forces between the fibers and nozzle walls align fibers along the flow direction. Simulation studies by Lewicki et al. [83] and Yang et al. [84] have shown that the nozzle's internal surface-to-volume ratio and the rheological properties of the polymer medium are critical factors for achieving effective shear alignment during extrusion. When these parameters are optimized, fiber alignment naturally occurs during fabrication.

The ongoing research focuses on optimizing process parameters and developing advanced strategies to improve fiber alignment in additive manufacturing. Achieving a high degree of fiber alignment (>90%) and precisely controlling fiber orientation in desired directions could significantly enhance the mechanical, thermal, and electrical properties of polymer composites. These advancements would have a transformative impact on the additive manufacturing of fiber-reinforced materials.

3.5 Additive manufacturing of continuous fiber-reinforced composites

Reinforcing composites with short fibers can significantly enhance strength, stiffness, and fracture resistance, provided the fibers are well-aligned in the printing direction and high fiber volume fractions are achieved. As shown in eq. (3.5), the improvement in strength and stiffness increases with the fiber aspect ratio (s). However, for applications requiring very high strength, such as those comparable to metallic parts, continuous fiber reinforcement must be considered.

Recently, the FFF technique has been employed by various researchers to produce continuous fiber-reinforced polymer composites. These studies utilized different fiber types, including glass, carbon, and Kevlar, to reinforce thermoplastic nylon using commercially available Markforged 3D printers [85, 86]. Results showed that the strength and modulus of composite samples increased significantly with higher fiber loading. The maximum improvements in strength and modulus were observed in the longitudinal (0°) direction, where fibers were deposited. Cross-hatch deposition (±45°) of fibers yielded the highest elongation at break for the same fiber loading [85].

Figure 3.7 illustrates four types of dog bone specimens fabricated for mechanical characterization: unreinforced nylon, carbon fiber composite with uniaxial reinforcement, Kevlar-reinforced composite with uniaxial fiber orientation, and Kevlar-reinforced composite with a cross-hatch (±45°) pattern.

To estimate the elastic modulus of continuous fiber-reinforced composites along the fiber direction, the rule of mixtures is a commonly applied method. The elastic modulus prediction using the rule of mixtures is expressed as follows:

$$E_c = E_r f + (1-f)E_m \qquad (3.9)$$

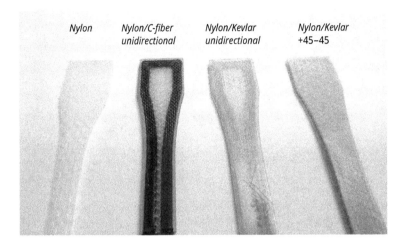

Fig. 3.7: 3D-printed continuous fiber-reinforced composites. Unreinforced nylon and composites reinforced with uniaxial carbon fibers, uniaxial Kevlar fiber, and crosshatched Kevlar fiber. The matrix material is nylon in all composites.

where E and f represent the elastic modulus and fiber volume fraction, respectively. Indices r, m, and c refer to the fiber reinforcement, matrix, and composite. Similarly, composite strength (σ_c) can also be predicted from the rule of mixture equation:

$$\sigma_c = \sigma_r f + (1 - f)\sigma_m \qquad (3.10)$$

According to this model, the strength and modulus of additively manufactured composites increase significantly with higher fiber volume fractions. High fiber volume fractions (approximately 40%) can be achieved using dual-nozzle FFF systems. In these systems, the first nozzle deposits the fiber reinforcement while the second nozzle extrudes thermoplastic polymer, which immediately solidifies and encapsulates the fibers. Alternatively, in-nozzle impregnation methods have been developed recently, where fiber bundles and thermoplastic melts are combined within the extruder and deposited through a single nozzle. This method also achieves high fiber volume fractions, as demonstrated in studies [87, 88]. A schematic of the in-nozzle impregnation process is presented in Fig. 3.8, illustrating its operational efficiency.

Continuous fiber reinforcement provides significant improvements in mechanical properties compared to discontinuous fibers. However, the additive manufacturing of continuous fiber composite systems presents additional challenges compared to powder and short fiber-reinforced composites. These challenges include managing dual-material feeding, slicing for multimaterial systems, and addressing weak bonding between fibers and the thermoplastic polymer. A previous study demonstrated that surface modification of carbon fiber bundle with methylene dichloride and PLA particles [87] improved adhesion and increased tensile and flexural strength. Oztan et al. [85] also reported that microstructural defects in the FFF process can significantly reduce

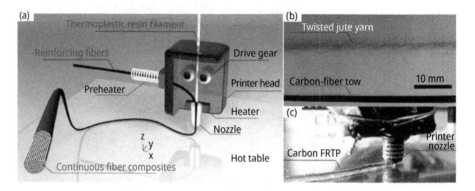

Fig. 3.8: FFF printing of continuous fiber composites by in-nozzle impregnation: (a) schematic; (b) fiber bundles used in FFF; and (c) photograph of the 3D printing process. Figure was reprinted with permission from [88].

the strength and stiffness of the composites. Research has demonstrated that the optimizing process parameters (i.e., print temperature, layer thickness, hatch spacing, and the number of fiber layers) as well as applying postprocessing techniques like thermal and pressure treatments can reduce defects and enhance the mechanical performance of these composites [86, 89].

As discussed in the section on short fiber composites, thermoplastic polymers tend to weakly adhere to reinforcing fibers. To address this limitation, continuous fiber-reinforced, thermoset composites have recently been introduced in additive manufacturing. Hao et al. [90] demonstrated a method in which a carbon fiber bundle is pulled through an epoxy resin pool, which then extruded onto the print bed through a 2 mm nozzle. This approach achieved a tensile strength of 792.87 MPa and an elastic modulus of 161.4 GPa. Similarly, Ming et al. [91] utilized a modified process with an initial impregnation step, as illustrated in Fig. 3.9A. In their study, a 3K carbon fiber tow was impregnated with E-20 epoxy resin at 130 °C, where the reduced resin viscosity facilitated penetration into the fibers. The impregnated tow was then fed into a heated printer head, passing through a viscous epoxy resin at 130 °C (similar to the wetting step) before being deposited onto the print surface (Fig. 3.9B). Upon cooling, the resin became more viscous, enhancing the shape stability. Following curing at 160 °C, the composite achieved an exceptionally high tensile strength of 1,476.11 MPa and a modulus of 100.28 GPa, setting a new benchmark for such materials.

Additive manufacturing of continuous fiber-reinforced thermoset composites is still in its early stages, and further research is needed to achieve the fabrication of composite structures with high mechanical performance. The proof-of-concept studies discussed above highlight the challenges and limitations associated with this emerging technology. However, it is evident that this area of research is poised for rapid growth in the near future, driven by the exceptional mechanical properties demonstrated in recent studies of additively manufactured continuous fiber-reinforced composites.

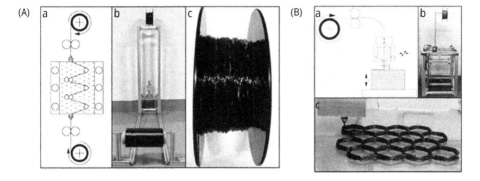

Fig. 3.9: Additive manufacturing of continuous carbon fiber-reinforced thermoset composites. Figure was reprinted from [91].

3.6 Mechanical performance comparison of additively manufactured polymer composites

Polymer composites can be additively manufactured using various types of reinforcements. Micro/nanosized dopants in powder form can be used to modify the thermal and electrical properties of polymers. Additive manufacturing techniques such as vat polymerization, extrusion, and material jetting can be employed to fabricate these composites. Similar methods are also used to produce SFRPCs, which enhance strength and stiffness, impact resistance, and reduce warpage. However, continuous fiber-reinforced composites are preferred for applications requiring higher strength. Currently, extrusion-based additive manufacturing is the only suitable method for fabricating continuous fiber composites. While their production is more challenging than that of discontinuous fiber-reinforced composites, continuous fiber composites offer significantly superior mechanical performance. Figure 3.10 illustrates a comparison of the strength and elastic modulus properties from selected studies on additively manufactured short and continuous fiber composites.

As illustrated in Fig. 3.10, continuous fiber-reinforced composites are significantly stronger and stiffer than short fiber composites. Additionally, thermoset composites demonstrate superior mechanical performance compared to thermoplastic matrix composites. However, there is a limited body of research on the additive manufacturing of thermoset-based composite materials. The figure also shows how the tensile strength of polymer composites varies with fiber volume fraction and the type of polymer matrix. Short and continuous fiber composites are categorized under two main groups: thermoset and thermoplastic composites. From the chart, it is clear that thermoset composites provide higher strength for the same fiber volume fraction compared to thermoplastic composites. This is primarily due to the weak adhesion between fibers and the thermoplastic matrix, as previously discussed.

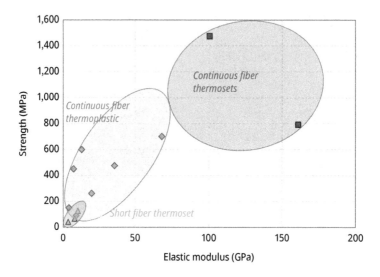

Fig. 3.10: Strength versus modulus comparison of additively manufactured fiber-reinforced composites.

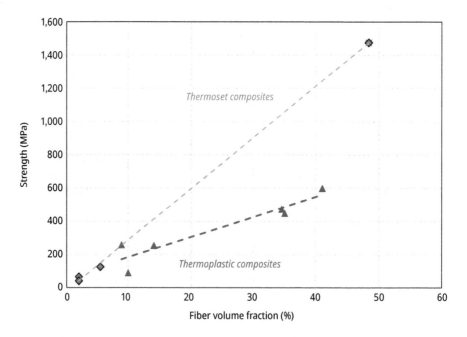

Fig. 3.11: Variation of strength as a function of fiber volume fraction.

Moreover, direct write additive manufacturing of thermoset composites results in fewer defects and less porosity compared to FFF process used for thermoplastic composites. As noted in the previous section and indicated in Fig. 3.11, short fiber-reinforced thermosets have significant potential for strength enhancement. However, current studies limit the fiber volume fraction to 5–6%. Moving forward, research efforts in composite additive manufacturing should focus on increasing the fiber volume fraction in short fiber-reinforced thermoset composites, improving fiber–matrix adhesion in thermoplastic composites, and optimizing process parameters for continuous fiber-reinforced thermoset composites.

4 Additive manufacturing of metals

Polymers and polymer composites currently dominate the feedstock materials used in additive manufacturing. However, additive manufacturing of metals is rapidly gaining popularity due to the superior mechanical performance and higher service temperatures metals offer compared to polymers. Additionally, the production of metal feedstock materials leverages established technologies, providing cost advantages. According to Wohlers Associates Inc. [92], a leading provider of strategic insights on additive manufacturing, sales of metal additive manufacturing systems experienced a remarkable ~80% growth in 2017 compared to 2016. This surge is driven by advancements in metal processing technologies, including improved part quality, faster fabrication speeds, and reduced costs of metal additive manufacturing systems.

Additively manufactured metals find applications across a wide range of industries, including dental, aerospace, biomedical, and automotive sectors. Biocompatible metals such as Ti-6Al-4V are widely used in orthopedics, enabling the fabrication of patient-specific implants. The ability to produce implants with controlled porosity provides advantages such as matching the mechanical properties of native joints and promoting bone ingrowth for enhanced functionality. In aerospace, additive manufacturing is revolutionizing the industry by enabling the production of lightweight, complex components that significantly reduce the overall weight.

In automotive industry, additive manufacturing is mainly preferred for rapid prototyping and the concept design stages rather than the end part manufacturing. Additive manufacturing, however, is preferred for manufacturing numerous components in luxury vehicles and automobiles that are customized for the user with limited number of part productions. In addition, for highly complex parts justified by a substantial vehicle improvement, additive manufacturing can be the preferred choice over the traditional manufacturing. The applications and adoption of additively manufactured metal components continue to expand across various industries, fueled by recent technological advancements.

As outlined in Chapter 1, various techniques are available for metal additive manufacturing, including powder bed fusion, direct energy deposition (DED), binder jetting, and extrusion. Figure 4.1 illustrates the market distribution of metal printing technologies utilized in 2019, as reported by Aniwaa [93].

DED employs metal feedstock in the form of powder or wire and is favored for fabricating large parts due to its high deposition speed. Wire-based DED is faster than powder-based methods, making it suitable for producing large components; however, its primary drawback is low resolution. Extrusion involves combining metal powders (typically up to 60–70%) with thermoplastic materials to produce filaments, which are then processed using the fused filament fabrication method. While this approach is simple and cost-effective, it does not achieve the high-grade metallic properties (mechanical, thermal, and electrical) characteristic of pure metals due to the metal–polymer compos-

https://doi.org/10.1515/9781501520242-004

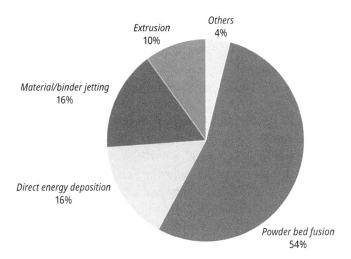

Fig. 4.1: Market breakdown in metal additive manufacturing technologies [94].

ite nature of the material. Binder jetting enables the production of high-resolution metal parts but involves multiple steps, including printing, binder removal, and sintering. Challenges with this method include component porosity and shrinkage after sintering, which can affect the final product's quality. Powder bed fusion, by contrast, is the most widely used metal additive manufacturing technology, accounting for more than half of the industry's market share, as illustrated in Fig. 4.1. This method is favored for its versatility, repeatability, minimal need for postprocessing, and compatibility with a wide range of powder feedstock. Consequently, the remainder of this chapter will focus on powder bed fusion, exploring its current status and advancements.

Powder bed fusion technologies are categorized into two main groups based on the energy source used to coalesce metal powders: laser-assisted manufacturing and electron beam melting. Laser-assisted manufacturing employs a high-power laser to partially or fully melt the powder feedstock to form the desired part. In contrast, electron beam melting utilizes an electron beam energy source in a vacuum environment to achieve layer-by-layer deposition. Regardless of the energy source, powder bed fusion requires high-quality metal powder feedstock to produce components with the desired performance. The next section discusses the common methods used to prepare metallic powder feedstock for the powder bed fusion process.

4.1 Feedstock material fabrication for powder bed fusion

The quality and performance of additively manufactured components heavily depend on the quality and cost of the metal powders used in powder bed fusion processes. These powders must meet strict requirements for chemical composition and morphology. Furthermore, the microstructural composition significantly influences the physical properties of the resulting manufactured parts. In general, metallic powders must be free from contamination to ensure high-quality outputs. Contaminant limits for metallic powders used in additive manufacturing are defined in ASTM standards F42.05 on Materials and Processes [95]. For instance, ASTM standards specify that the oxygen and nitrogen content in finished Ti-6Al-4V products must not exceed 0.2 wt% and 0.05 wt%, respectively [96]. Additionally, powder bed fusion methods demand powders with excellent flowability, which requires a spherical particle shape and reasonably large particle sizes. However, while smaller particles reduce flowability, larger particles can negatively affect geometric resolution and surface quality. Therefore, spherical metallic powders with low contamination and good flowability are the preferred feedstock materials for metal additive manufacturing.

To produce spherical metallic powders suitable for powder bed fusion, four primary processing routes are commonly used: water atomization, gas atomization (GA), plasma atomization, and the plasma rotating electrode process (PREP). Atomization in powder manufacturing refers to the process of breaking down bulk liquid metal into small droplets, which then solidify into spherical particles upon cooling.

In water atomization, the liquid alloy free falls through an atomization chamber and is rapidly cooled and atomized by the surrounding water jets. As the metal droplets solidify, the fabricated powder is collected at the bottom of the chamber and dried. Due to the high cooling rates, powders produced through this method are typically coarse and irregular in shape, which reduces both packing density and flowability properties [97]. Furthermore, oxidation is a significant concern in water atomization, as it affects powder flow behavior, alters melt pool dynamics, and ultimately impacts the bulk material composition and mechanical properties of the final part. Consequently, water atomization is unsuitable for reactive materials like titanium and is primarily used for producing steel powders.

GA addresses the limitations of water atomization by using high-pressure gas jets, rather than water, to atomize the liquid metal (Fig. 4.2A). Inert gases such as nitrogen or argon are typically employed to minimize oxidation and contamination. Due to the lower heat capacity of gas compared to water, the metal droplets cool more slowly, resulting in more spherical particles with improved flowability. However, contamination risks remain due to potential interactions with ceramic crucibles and atomizing nozzles used in the process [97]. To mitigate contamination further, electrode induction melting GA (EIGA) can be utilized. In the EIGA process, a metal bar is melted by an induction coil before entering the atomization chamber, where the molten metal flows directly into the gas stream for atomization. Since the metal feedstock does not contact the crucible or electrodes, contamination is significantly reduced. Despite these improvements, internal voids within individual particles caused by trapped argon remain a concern for gas-atomized powders [98].

Plasma atomization is similar to the EIGA process in that it uses a metal wire as the feedstock, as illustrated in Fig. 4.2C. However, unlike EIGA, where induction coils are used,

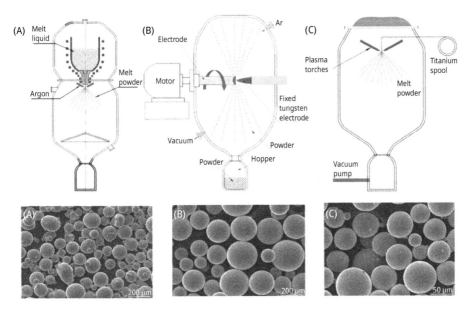

Fig. 4.2: Schematic of atomization methods and SEM images of Ti6Al4V powders fabricated by (a) gas atomization, (b) plasma rotating electrode process, and (c) plasma atomization. Reprinted with permission from [99].

plasma atomization employs plasma torches to melt the metal wire in a hot zone. The molten wire is then broken into fine droplets, which cool rapidly. Like EIGA, plasma atomization produces high-purity metal powders since the molten metal does not come into contact with crucibles or electrodes that could introduce contaminants before solidification.

One advantage of plasma atomization is its significantly higher yield of fine powder compared to conventional GA. The resulting powders exhibit excellent sphericity and fewer satellite particles than gas-atomized powders. However, issues such as porosity caused by trapped gas during atomization still persist in plasma-atomized powders [98]. Additionally, the process requires wire feedstock, which increases fabrication costs and limits its applicability to a narrower range of metals and alloys.

A newer variant of plasma atomization, known as plasma spheroidization, uses metal powders as feedstock instead of wires. This method does not alter the particle size but converts irregularly shaped particles into spherical powders. The resulting powders have high sphericity, with impurity levels dependent on the quality of the initial feedstock. Using powder feedstock and producing minimal satellite particles are significant advantages of this technique. However, the high investment costs associated with plasma atomization remain a drawback.

For producing spherical metal powders with the highest purity, the PREP is preferred. In this technique, a rotating metal rod electrode is melted by a tungsten-tipped cathode. The molten metal is ejected from the electrode's surface by the centrifugal force, forming droplets that solidify in-flight into spherical powders. PREP powders are highly pure due to the absence of contact with external materials during the process and exhibit minimal gas porosity compared to other methods relying on high-pressure gas. Additionally, the centrifugal force prevents droplet collisions, reducing the formation of satellite particles [98].

Despite these advantages, PREP has some drawbacks, including high costs and low powder yields. The process typically produces powders that are too coarse for powder bed fusion applications. Increasing the rotation speed of the electrode can reduce particle size but may introduce dynamic instability and balancing issues, as noted in [100]. A comparison of these powder manufacturing techniques is summarized in Tab. 4.1, highlighting the key properties of each method.

Tab. 4.1: Summary of metal powder processing technologies (modified from [97]).

Manufacturing process	Particle size (µm)	Advantages	Disadvantages	Metal types
Water atomization	10–500	High throughput Range of particle sizes Feedstock in ingot form	Low sphericity Satellites present Wide PSD Low yield in 20–150 µm	Nonreactive
Gas atomization	10–300	Wide range of materials Feedstock in ingot form High throughput High sphericity	Satellites present Wide PSD Low yield in 20–150 µm	Ni, Co, Fe, and Al Ti (EIGA)
Plasma atomization	10–200	High purity High sphericity	Feedstock in wire or powder form High cost	Ti and Ti-6Al-4V
PREP	10–100	Highest purity High sphericity	Very low yield High cost	Ti and exotics

New powder production methods are being developed to achieve high-purity, low-cost feedstock fabrication. Techniques such as electrolytic processes, metallothermic methods, and the hydride–dehydride process show significant potential as cost-effective alternatives for metal powder production [101].

4.2 Feedstock materials used in metal additive manufacturing

Common materials used in metal additive manufacturing include steel (stainless and tool steels), aluminum, titanium and its alloys, as well as nickel-based alloys (e.g., Inconel) and other metallic materials. Figure 4.3 illustrates the usage percentages of these metals in the additive manufacturing industry. Currently, stainless steel and titanium-based materials dominate the metallic materials utilized in additive manufacturing applications.

Steel, as one of the most widely used engineering materials, also holds a leading position in metal additive manufacturing, as shown in Fig. 4.3. Its excellent mechanical properties make it a preferred choice for a variety of applications. Additionally, metal additive manufacturing technologies are well-suited for fabricating a range of steel types. Among these, austenitic stainless steels (e.g., 316L) are the most commonly used, although maraging steels, precipitation-hardenable stainless steels, martensitic cutlery-grade steels, and tool steels (e.g., H13) have also been extensively explored.

The mechanical properties of steels are highly dependent on their microstructure, which can be tailored through precise control of process parameters such as cooling rates, temperature gradients, and elemental composition. Metal additive manufacturing provides the capability to produce metallic parts with unique microstructures and customized mechanical properties by optimizing these parameters during fabrication.

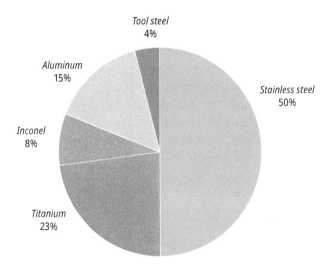

Fig. 4.3: Commonly used metals in powder bed fusion additive manufacturing [102].

4.2.1 Titanium and titanium alloys

Titanium and its alloys are increasingly popular in additive manufacturing due to their exceptional strength, lightweight properties, and biocompatibility. These attributes have enabled their widespread adoption in industries such as automotive, aerospace, and healthcare. Additionally, the high cost of titanium materials and the challenges associated with conventional manufacturing methods make them particularly well-suited for specialized additive manufacturing applications.

Pure titanium and the Ti-6Al-4V alloy are among the most commonly used materials in commercially available powder bed fusion technologies. Although less frequently other titanium alloys have also been explored for specific applications, for instance, Ti-24Nb-4Zr-8Sn and Ti-6Al-7Nb are being investigated for biomedical uses, while Ti-6.5Al-3.5Mo-1.5Zr-0.3Si has shown promise in aerospace applications [101].

4.2.2 Aluminum alloys

Unlike steel and titanium, aluminum is relatively easy and cost-effective to manufacture using conventional methods such as casting and machining. Consequently, the commercial advantage of producing aluminum parts through additive manufacturing is less significant. Additionally, aluminum alloys pose challenges in additive manufacturing due to their poor flowability during the recoating process in powder bed fusion and their high reflectivity during laser-based fusion processes [101]. These manufacturing difficulties, coupled with lower commercial incentives, have limited the widespread use of aluminum in additive manufacturing.

Despite these challenges, aluminum and its alloys remain attractive for additive manufacturing due to their low cost and lightweight properties. Aluminum's high thermal conductivity helps reduce thermally induced stresses, minimizing the need for extensive support structures, and also enables higher processing speeds. With the advancement of new additive manufacturing technologies that address these challenges and the growing availability of aluminum powder suppliers, aluminum is expected to become a key material for additive manufacturing, particularly as the industry moves toward larger-scale production of mass goods [103].

4.2.3 Other metals

High-performance metals beyond steel, aluminum, and titanium have also garnered significant attention within the additive manufacturing community. Among these, nickel-based superalloys are prominent for high-temperature applications, with materials such as Inconel 625, Inconel 718, and NiCoCr also finding applications in the biomedical field [101]. Additionally, Invar (nickel–iron) alloys are utilized in applications requiring a low coefficient of thermal expansion, particularly in powder bed fusion processes.

High-entropy alloys (HEAs) are another category of advanced materials that have drawn considerable interest. These alloys are defined by their composition of three or more principal elements in equal or nearly equal atomic percentages [104]. HEAs are single-phase solid–solution alloys that exhibit unique mechanical and chemical properties, including a combination of high yield strength and ductility, microstructural stability, exceptional mechanical performance at elevated temperatures, and high resistance to corrosion and oxidation. As a result, HEAs have become a focal point for additive manufacturing research. Powder bed fusion techniques have successfully fabricated various HEAs, including FeCoCrNi, AlCoCrFeNi, and CoCrFeMnNi [104–106]. Studies on these materials report superior strength and ductility, attributed to the formation of a single-phase solid–solution structure.

4.3 Design considerations in metal additive manufacturing

Unlike conventional manufacturing, additive manufacturing of metals requires careful consideration of several process-related factors. These include void formation, residual stresses, surface roughness, and postprocessing requirements. Each of these parameters can have a significant impact on the mechanical performance of the fabricated parts and must therefore be thoroughly evaluated and optimized. The following section discusses these process parameters and their effects on the performance of additively manufactured metal components.

4.3.1 Void formation

Spatter ejection, also known as powder jump, occurs when metal powder flies out of the laser's path and settles on parts of the fabricated material. This contamination of the powder bed can compromise the quality of each printed layer and is more pronounced at high laser intensities. In addition to spatter, high laser intensity can increase melt pool dynamics and reduce part density due to the formation of pores from entrapped gas [101]. These pores are typically spherical, as they form during the evaporation of material.

Conversely, if the laser intensity is too low, the metal powders may not fuse properly, resulting in loose powders that create voids within the printed part. Voids caused by unmolten material tend to have irregular shapes, which lead to higher stress concentrations compared to spherical pores. These irregular voids act as initiation sites for cracks during mechanical testing. Therefore, irregularly shaped voids are more detrimental to the mechanical properties of the material than spherical pores, as they can cause premature material failure.

4.3.2 Residual thermal stresses

In metal 3D printing, pristine alloy feedstock undergoes rapid melting and solidification during processing, which can induce thermal stress in the finished part [107]. If these residual stresses exceed the material's strength or that of the substrate, defects such as part cracking or substrate warping may occur. Minimizing residual stress can be achieved through postprocess heat treatment and by optimizing printing process parameters. Thermal stresses are also material-dependent, with different metals – such as aluminum, steel, and titanium – exhibiting varying sensitivities to volumetric changes under identical thermal conditions. This variability results in differing levels of residual thermal stress among materials.

4.3.3 Surface roughness

Surface roughness is a defect that impacts the quality, appearance, and mechanical performance of 3D-printed parts. While surface roughness may not significantly influence the static mechanical properties of additively manufactured components, it can be detrimental under fatigue loading conditions, as will be discussed later. As a result, it is common practice to reduce surface roughness through mechanical or chemical postprocessing methods. Factors such as powder size, cooling rate, and spatter can all influence the surface roughness of the printed parts.

4.3.4 Postprocessing

Postprocessing is often applied to fabricated metallic components to alleviate thermal stresses, reduce porosity, optimize grain microstructure, and improve surface roughness. Unlike polymer additive manufacturing, postprocessing procedures for metal components are typically carried out, while the parts remain attached to the build platform. Recently, ASTM has established standardized postprocessing conditions for metals produced through powder bed fusion in the ASTM F3301 standard [108].

4.3.4.1 Stress relief
As mentioned earlier, residual stresses can negatively impact both the static and dynamic mechanical properties of additively manufactured metallic components. These stresses must be relieved before the part is removed from the build plate to prevent warping or cracking. Stress relief is typically achieved by placing the entire build plate in an oven. The temperature and duration of the stress-relief process are specified for metals in the ASTM F3301 standard.

4.3.4.2 Heat treatment
In addition to stress relieving, heat treatments such as aging and solution annealing are applied to optimize the grain microstructure and enhance the mechanical properties of the parts. Heat treatment can cause dimensional distortions, so it is generally performed before the final surface finishing and machining steps.

4.3.4.3 Hot isostatic pressing

Hot isostatic pressing (HIP) is a process that involves applying both high temperature and pressure simultaneously to reduce porosity in metallic components. A furnace within a pressure vessel controls the temperature, while inert gas is used to apply uniform isostatic pressure. This pressure causes internal voids to collapse, leading to densification of the material. By eliminating defects and voids, the HIP process significantly improves the mechanical properties of the components. It is a well-established technique for a wide range of materials, including titanium, steel, aluminum, and superalloys. In additive manufacturing, HIP is commonly used to achieve near-theoretical densities by completely eliminating porosity in the parts.

4.3.4.4 Machining and surface treatments

Mechanical machining may be necessary to ensure the dimensional accuracy of fabricated metal parts. Additionally, surface finishing may be required to improve surface quality, reduce roughness, clean internal channels, or remove partially melted particles. As noted earlier, surface roughness not only detracts from the aesthetics of the part but also significantly lowers its fatigue performance. Surfaces can also undergo chemical treatments or coatings to enhance physical appearance and improve functional properties, such as reactivity, biocompatibility, corrosion resistance, and wear resistance.

4.4 Mechanical properties of additively manufactured metals

As previously mentioned, there is a strong correlation between the process-dependent microstructure and the mechanical properties of additively manufactured metals. Figure 4.4 illustrates the relationship between the additive manufacturing process parameters, microstructure, and the mechanical performance of the fabricated parts. Process parameters such as laser intensity, scan speed and direction, powder type and morphology, and postprocessing treatments all influence the final microstructure. Key microstructural properties in additively manufactured metals include grain size and orientation, porosity, roughness, and density. These microstructural characteristics, shaped by the selected process parameters, directly determine the mechanical properties of the parts. Key mechanical properties, including strength (tensile and yield), hardness, elongation at break, and fatigue limit, define the structural performance of the fabricated components.

Several metallurgical processes are crucial in achieving the ideal microstructure in additively manufactured parts for specific applications. Processes such as grain refinement, precipitation hardening, solid–solution strengthening, and age hardening play significant roles in defining the final microstructure and, consequently, the mechanical properties of the fabricated metal. Selecting the appropriate elemental composition of the metal feedstock and controlling both process and postprocess condi-

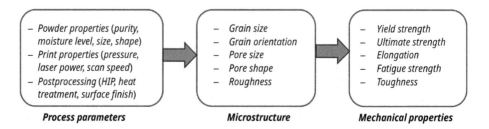

| Process parameters | Microstructure | Mechanical properties |

- Powder properties (purity, moisture level, size, shape)
- Print properties (pressure, laser power, scan speed)
- Postprocessing (HIP, heat treatment, surface finish)

- Grain size
- Grain orientation
- Pore size
- Pore shape
- Roughness

- Yield strength
- Ultimate strength
- Elongation
- Fatigue strength
- Toughness

Fig. 4.4: Relationship between the process parameters, microstructure, and mechanical properties of additively manufactured metals.

tions are active areas of research, though they fall outside the scope of this book. Extensive studies have been conducted to better understand the strengthening mechanisms in metals [109, 110] and the relationship between microstructure and mechanical properties in additively manufactured metallic components [101].

Porosity is a key factor influencing the mechanical performance of metal structures. As discussed earlier, pores (or voids) facilitate crack initiation and propagation, significantly compromising the mechanical performance of additively manufactured parts. Not only the total porosity but also the pore size distribution and pore shape influence the final mechanical properties. Therefore, minimizing porosity is a primary objective in metal additive manufacturing. Recent advancements in metal additive manufacturing technologies enable the production of metallic parts with densities exceeding 99.5% for various metals. This density can be further enhanced, and thermal stresses that arise during fabrication can be reduced through postprocessing operations like HIP and thermal annealing.

By minimizing porosity and optimizing additive manufacturing process parameters, it is possible to produce metallic parts with static mechanical strengths comparable to those fabricated using conventional methods. In fact, the static strength of additively manufactured parts can even exceed that of conventionally produced parts (e.g., castings) due to the finer grain structure achieved through additive manufacturing [101]. However, strength is typically anisotropic in additive manufacturing, as grains are preferentially oriented and elongated along the printing direction. As a result, higher strength is generally observed in the printing direction compared to the orthogonal directions. The static mechanical properties of selected metals are provided in Tab. 4.2.

As given in this table, the static strength of additively manufactured metals can match or even surpass that of their conventionally produced counterparts. However, the elongation at break and ductility of additively manufactured metals typically exhibit significant variation and are highly dependent on the process conditions and postprocessing treatments. Key factors that reduce ductility in these metals include residual stresses, oxygen content, and porosity [101]. Ductility can be improved, if nec-

Tab. 4.2: Static mechanical properties of selected metals fabricated via powder bed fusion (PBF) or conventional (cast and wrought) manufacturing (redrawn according to [101]).

Metal type	Process	Postprocess	Yield strength (MPa)	Ultimate tensile strength (MPa)	Elongation (%)	References
316L stainless steel	Wrought	Annealed	170	485	40	[111]
	PBF	No	444	567	8	[112]
	DED	No	410	640	36	[113]
304 Stainless steel	Wrought	Annealed	170	485	40	[111]
	LBM	No	182	393	25.9	[114]
17-4PH precipitation hardening steel	Wrought	Sol. treated and aged	1,170	1,310	10	[115]
	PBF	Sol. treated	939	1,188	9	[116]
18Ni-300 Maraging steel	PBF	No	1,214	1,290	13.3	[117]
H13 high-speed steel	DED	No	1,505	1,820	6	[118]
AlSi12	Cast	No	130	240	1	[119]
	PBF	No	260	380	3	[120]
AlSi10Mg	Cast	No	140	240	1	[119]
	PBF	No	230	328		[121]
AlMg1SiCu	PBF	HIP	–	42	–	[122]
ALMg4.4Sc0.66MnZr	PBF	Aged	520	530	14	[123]
Ti	Sheet	No	280	345	20	[124]
	PBF	No	555	757	19.5	[125]
Ti-6Al-4V	Cast	No	896	1,000	8	[126]
	Wrought	Sol. treated and aged	1,100	1,170	12	[127]
	PBF		1,040	1,140	8.2	[128]
	PBF	No	1,160	1,240	11.5	[129]
	DED	Heat treated (in situ) No	945	1,041	18.7	[130]
Ti-6.5Al-3.5Mo-1.5Zr-0.3Si	PBF	No	1,030	1,101	10.2	[131]

essary, through heat treatment, which helps alleviate thermal stresses and promote grain microstructure reorganization.

Fatigue strength is another critical property that determines the durability of metals under dynamic loading. Like static strength, the fatigue performance of metallic components is influenced by microstructural factors such as grain size and orientation, as well as defects and porosity. However, unlike static strength, fatigue strength is also highly dependent on surface finish and roughness. Surface defects, unmelted metal particles, and high surface roughness can create stress concentrations and initiate crack formation, significantly reducing the fatigue performance of the component. Surface roughness can be reduced through machining, sandblasting, or polishing, which can notably improve fatigue properties. Additionally, postprocessing methods such as HIP and thermal treatment can further enhance fatigue performance by reducing porosity and optimizing the grain microstructure, respectively. As given in Tab. 4.3, postprocessing operations can improve the fatigue strength of additively manufactured metals to levels comparable to those of cast and wrought materials.

In summary, metal additive manufacturing is gaining rapid popularity due to recent advancements in the technology and improved control over manufacturing process parameters. As a result, the mechanical performance of these materials can approach, and in some cases exceed, the static mechanical properties of conventionally processed materials. However, limitations such as weak fatigue performance, lower

Tab. 4.3: Fatigue strengths of selected metals fabricated via powder bed fusion (PBF) or conventional (cast and wrought) manufacturing (redrawn according to [101]).

Metal type	Process	Postprocess	Surface treatment	Fatigue strength at 10^7 cycles (MPa)	References
316L stainless steel	PBF	No	No	108	[132]
		Heat treated	Machined	294	[132]
		HIP	Machined	317	[132]
17-4PH precipitation hardening steel	PBF	No	Polished	407	[133]
AlSi12	Cast	No	Polished	55–80	[134]
	PBF	Stress relieved	Polished	80	[135]
AlSi10Mg	Cast	No	Polished	45	[134]
	PBF	No	No	70	[136]
	PBF	No	Polished	115	[136]
AlMg4.4Sc0.66MnZr	PBF	Aged	Polished	300	[137]
Ti-6Al-4V	Wrought	No	Polished	675	[138]
	PBF	Stress relieved	No	200	[139]
	PBF	Stress relieved	Polished	300	[139]
	PBF	Stress relieved	No	210	[140]
	PBF	Stress relieved	Polished	500	[140]

ductility, and high variability in results – depending on factors like the manufacturing process, equipment, feedstock type, and other parameters – restrict the broader adoption of these techniques for fracture-critical applications. Further research is needed to address the underlying causes of these issues and resolve them, making metal additive manufacturing a more viable alternative to traditional metal fabrication methods.

5 Additive manufacturing of ceramics

Ceramic materials exhibit unique mechanical, thermal, and electrical properties that are unmatched by polymers and metals. These include exceptional hardness, excellent thermal and electrical insulation, chemical and wear resistance, and extremely high melting points. Such characteristics make ceramics ideal for applications in harsh environments involving high wear, temperature, or pressure. Additionally, ceramics are abundant, naturally occurring materials that offer excellent biocompatibility and cost-effectiveness.

Traditional ceramics, derived from raw materials like clay and quartz sand, date back to ancient civilizations and have been extensively used in applications such as foodware, bricks, tiles, industrial abrasives, refractory linings, and Portland cement. Beyond these traditional uses, high-performance ceramics, such as oxides, carbides, and nitrides, have gained prominence for their superior mechanical properties, including high strength and resistance to high temperatures, corrosion, and oxidation. These materials are increasingly utilized in aerospace, automotive, and energy sectors for components like gas turbines, engines, batteries, and heat exchangers, which operate under extreme conditions. However, the fabrication of ceramics into complex geometries using conventional manufacturing methods is challenging due to their low ductility, high melting points, and sensitivity to cracking.

Additive manufacturing (AM) provides a transformative solution for producing ceramic materials in intricate geometries, enabling the fabrication of customized and lightweight components through topology optimization. This design flexibility has broadened the applications of additively manufactured ceramics. The biocompatibility of ceramics, combined with the ability to create patient-specific parts, has led to significant adoption in biomedical applications. Bioinert ceramics like alumina and zirconia are widely used in dental applications, while bioactive materials such as hydroxyapatite (HA) and bioactive glasses are favored for orthopedic implants like knee, hip, and spinal fusion devices, promoting bone tissue repair and growth. In aerospace, additively manufactured ceramics find applications in high-temperature-resistant and robust components, including turbine parts, spark plugs, and hypersonic engine elements, which are difficult or impossible to produce using traditional methods. Figure 5.1 showcases the examples of additively manufactured ceramics used in dental and jewelry applications.

Ceramic AM can be categorized into three primary types based on the form of ceramic feedstock: powder-based ceramic manufacturing, slurry-based ceramic manufacturing, and bulk solid-based ceramic manufacturing, as illustrated in Fig. 5.2. The processing methods, along with their respective advantages and disadvantages, are detailed below.

https://doi.org/10.1515/9781501520242-005

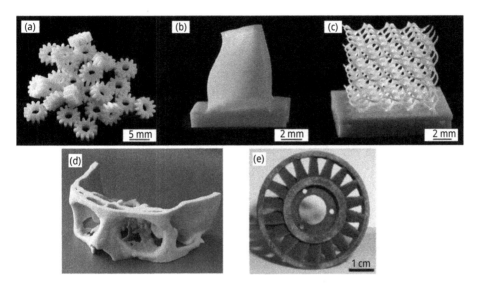

Fig. 5.1: Examples of additively manufactured ceramic parts used for different applications: (a) alumina gears parts fabricated using the LCM technique, (b) alumina turbine blade, and (c) alumina cellular cube. (a–c) were reprinted with permission from [141], (d) cranial segment fabricated via powder bed fusion (figure was reprinted with permission from [142]), and (e) SiOC-based turbine wheel (figure was published with permission from [143]).

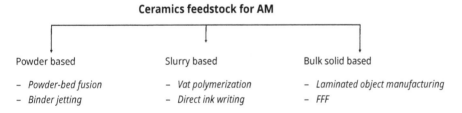

Fig. 5.2: Classification of additive manufacturing techniques based on the ceramic feedstock type.

5.1 Powder-based ceramic additive manufacturing

Powder-based ceramic AM utilizes loose ceramic powders as the feedstock material. In this method, the ceramic powders are either bonded by spraying a liquid binder (binder jetting) or melted (partially or fully) using a laser beam (powder bed fusion). Both binder jetting and powder bed fusion are widely used in ceramic AM due to their design flexibility and the ability to fabricate parts without requiring support structures.

5.1.1 Binder jetting of ceramics

Binder jetting is the most widely used AM technology for ceramics. As outlined in Chapter 1, this process employs an inkjet printing head to deposit a binding liquid onto ceramic powder, shaping it into a three-dimensional form. After printing, the binder is removed, and the remaining ceramic particles are sintered to enhance the stability and strength of the fabricated part. Multiple nozzles can be utilized to produce large objects (up to meters long) or to simultaneously create layers for smaller objects in mass production scenarios [144].

Binder jetting offers notable advantages including high speed, low cost, and design flexibility. This technique is particularly suitable for fabricating ceramics with high porosity, which is advantageous for specific applications. Porous biomedical ceramics are commonly used in tissue engineering for components such as scaffolds, implants, and cages, where the porous network supports tissue and cell ingrowth. Materials like HA and calcium phosphates are often employed in binder jetting to create scaffolds and implants for bone replacement.

While porosity benefits scaffolding applications, it can adversely affect the mechanical performance of ceramic materials. Porosity arises from the imperfect packing of ceramic particles and the sacrificial nature of the binder, which is removed during sintering. Porosities ranging from 40% to 60% are typical in binder jetting. Unlike metals, ceramics are more sensitive to defects and have low fracture toughness, making porosity a significant concern for mechanical stability.

To address this issue, various postprocessing techniques have been developed to enhance density and structural integrity. These methods include infiltration, isostatic compaction before sintering, and liquid-phase sintering.

Infiltration: In this process, a secondary liquid material, such as molten metal or ceramic/metal slurries, is introduced into the porous ceramic part. This fills the voids and densifies the structure, significantly improving mechanical properties by reducing porosity. In some cases, the infiltrated material reacts with the porous preform to create new ceramic phases. For instance, reaction-bonded silicon carbide ceramics can be produced by infiltrating liquid silicon into binder-jetted preforms and heating them, forming silicon carbide [145, 146] or Ti_3SiC_2 [147] depending on the preform material.

Isostatic compaction: Applying pressure to "green" parts before sintering can significantly reduce porosity. Cold isostatic pressing has been shown to produce highly dense (99%) ceramic structures, such as Ti_3SiC_2 [148]. Similarly, warm isostatic pressing, which applies heat and pressure simultaneously, has been used to achieve high densities, as demonstrated by Yoo et al., who achieved 99.2% relative density in alumina parts [149]. However, this method is less effective for parts with complex geometries, as internal cavities may deform or fail under pressure.

Liquid-phase sintering: Adding secondary phases or materials to the ceramic mix prior to printing can enhance densification during sintering. At elevated temperatures, these additives form a viscous liquid that improves bonding and fills cavities. This method has yielded highly dense (95–98%) ceramics, such as tricalcium phosphate (TCP) with ZnO/SiO_2 additives [150] and glass ceramics with HA/wollastonite additives [151].

In summary, binder jetting is a versatile and widely adopted technology for producing porous ceramic structures with complex geometries. Such ceramics can often be used directly without additional processing. However, achieving high-density ceramic components comparable to bulk theoretical densities requires postprocessing techniques like infiltration, isostatic compaction, or liquid-phase sintering. These methods enable the production of ceramic parts with enhanced mechanical properties, suitable for demanding applications.

5.1.2 Powder bed fusion of ceramics

Powder bed fusion shares similarities with binder jetting in that it does not require additional support material during the printing process; the powder bed itself serves as support for successive layers. However, unlike binder jetting, powder adhesion and material consolidation in this method are achieved by melting loose powders with a laser beam rather than by applying an adhesive binder.

Direct sintering of ceramic powders using a laser beam is challenging due to ceramics' low thermal shock resistance, poor sintering characteristics, and the high costs associated with their extremely high melting temperatures. Consequently, direct melting of ceramic powders has been explored in only a limited number of studies. In these studies, alumina [152], zirconia [153], and zirconia/yttrium oxide (ZrO_2–Y_2O_3) [154] powders were used to fabricate parts. However, the resulting components exhibited high porosity, crack formation, and rough surface finishes. To address these issues, the laser microsintering (LMS) technique was developed [155], which enables ceramic powder sintering at the submicron scale with high resolution and reduced surface roughness. Despite its advantages, LMS is only suitable for small parts (microscale to millimeter scale) and is impractical for larger components.

To overcome the challenges of direct ceramic powder sintering, an alternative powder bed fusion approach has been developed. In this method, ceramic powders are mixed with a secondary material, such as a lower melting temperature binder, which melts during the laser process to form the desired 3D shape. These binders are selected for their thermal shock tolerance to reduce residual stresses and cracking during sintering. Binder materials can include polymers, metals, or ceramics (e.g., glass).

If an organic polymer binder is used, it is removed after laser sintering by heating, and the remaining ceramic powder is subsequently sintered in a furnace at high temperatures. In contrast, inorganic binders are not removable by firing and either remain in the final material or interact with the ceramic powder to form secondary phases. For metallic binders, fabrication is conducted under an inert gas atmosphere to prevent oxidation. The binder can either be mixed with the ceramic powder or coated onto the powder surface, with the latter method offering more homogeneous distribution, improved sintering, and better mechanical properties.

Various binders have been employed in powder bed fusion including PEEK, polyamide, ammonium phosphate, TCP, polycarbonate, glass, and boron oxide. The use of sacrificial binders typically results in significant porosity, making these materials suitable for biomedical applications where porosity is advantageous. However, for structural applications requiring dense ceramics, additional densification methods, such as infiltration and isostatic compaction (discussed in the previous section), are applied to porous ceramic preforms. Despite its benefits, the organic binder-based powder bed fusion process has limitations. It cannot be used for enclosed geometries, as the binder cannot be adequately removed from such structures. Additionally, this process often results in low geometric resolution and significant shrinkage during sintering and compaction stages, posing further challenges for achieving precise dimensions and properties in fabricated ceramics.

5.2 Slurry-based ceramic additive manufacturing

Slurry-based ceramic AM involves preparing a liquid feedstock, typically in the form of an ink or paste, where ceramic particles are dispersed. This ceramic slurry can then be used in AM processes such as vat polymerization, material jetting, or paste extrusion.

5.2.1 Vat polymerization of ceramics

In vat polymerization, ceramic particles are embedded within a photopolymer resin. As the photopolymer cures under a UV light source, the ceramic material becomes encapsulated within the cross-linked polymer network. The fabricated part typically undergoes postprocessing steps, including heat treatment, to remove the polymer phase and sinter the ceramic particles.

Achieving high-quality parts requires careful control of the slurry's homogeneity, stability, and viscosity. Ceramic particles must be well-dispersed throughout the photopolymer, with stability maintained during the printing process. Poor dispersion or particle segregation can lead to inhomogeneous parts with suboptimal performance. While a lower ceramic content improves dispersion, it results in a porous ceramic network and significant shrinkage, which are undesirable for structural applications. Viscosity is another critical factor; the ceramic–photopolymer mixture must exhibit low viscosity and smooth flow characteristics to suit the vat polymerization process. High ceramic content can also hinder the photopolymerization process, as ceramic particles may absorb or scatter UV light, causing incomplete curing.

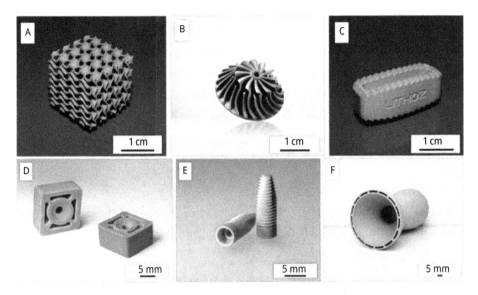

Fig. 5.3: Images of a silicon nitride-based ceramic part fabricated via LCM additive manufacturing: (A) gyroids, (B) impeller, (C) spinal implant (PLIF cage), (D) cutting tools, (E) dental two-piece implants with M1.6 inner thread, and (F) de Laval nozzle (figure was reprinted from [157]).

Various ceramics, including SiO_2, Al_2O_3, ZrO_2, and SiC, have been successfully fabricated using vat polymerization. Stereolithography and digital light processing (DLP) are the primary vat polymerization technologies employed for ceramic AM. As described earlier, DLP projects an entire 2D light pattern rather than scanning point-by-point, significantly increasing printing speed. A notable commercial application of this technology is lithography-based ceramic manufacturing (LCM), pioneered by Lithoz GmbH [156]. Figure 5.3 illustrates examples of ceramic components produced using LCM polymerization.

Due to the speed and the resolution of the process, fine ceramic structures including lattice structures [158], heat exchangers [159] and negative Poisson's ratio metamaterial structures [160] have been fabricated with this ceramic additive fabrication method.

5.2.2 Direct writing (DW) of ceramics

Direct ink writing (DIW) is an AM technique in which viscous liquids or pastes are extruded layer by layer to create 3D objects. This method offers a faster and more cost-effective alternative to photopolymerization. Ceramic suspensions used in DIW typically consist of ceramic powders, dispersants, polymer binders (such as polyvinylpyrrolidone or methyl cellulose), and solvents like water. Alumina is a common material choice for this technique due to its low cost, wide availability, and ease of densification.

Highly loaded viscous inks, containing over 60% ceramic content, can be prepared to minimize shrinkage during the sintering process [161]. Similar to other ceramic AM methods, DIW has been widely used for biomedical applications, particularly for fabricating porous ceramic structures that mimic the architecture of natural bone. Calcium phosphate glasses and HA are frequently chosen materials for DIW due to their exceptional biocompatibility, making them ideal for artificial bone scaffolds.

Fig. 5.4: PZT structure fabricated via direct ink writing: (A) low magnification image of the porous PZT structure; (B) SEM image showing detailed microstructure (images were reprinted from [162]).

Figure 5.4 shows low- and high-magnification images of a lead zirconate titanate (PZT) ceramic component fabricated via DIW [162]. PZT is a piezoelectric ceramic known for its ability to deform in response to an electric field, making it valuable in a variety of applications. A recent study [163] utilized DIW to fabricate bioinspired ceramic materials that could not be produced using other AM techniques. In this research, hexagonal alumina platelets (~5 μm in diameter, 0.5 μm in thickness) were mixed with submicron alumina powder and 3D printed into various geometries. The high shear stresses generated during the direct write process caused the alignment of the alumina platelets perpendicular to the nozzle wall, resulting in concentric layers of platelets, as illustrated in Fig. 5.5. The alignment was influenced by the shear forces, which were strongest near the nozzle wall, causing the platelets to align well in that region, while the platelets were more randomly oriented toward the core where the yield stress was lower. The study found that the nozzle length was a key factor in platelet alignment. Longer nozzles provided more time for platelet alignment, leading to a reduced core size with more randomly oriented platelets.

In addition to controlling the alignment of alumina nanoplatelets, this study also successfully fabricated bioinspired Bouligand mesostructures. A Bouligand structure is a helical arrangement of successive layers, as depicted in Fig. 5.6. This structural pattern is commonly found in natural materials such as plywood and chitin-protein fibers in

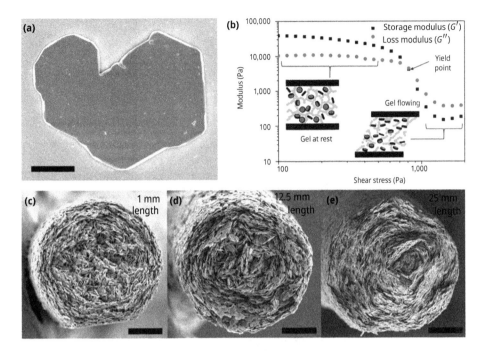

Fig. 5.5: (a) SEM image of a single alumina platelet; (b) dynamic mechanical analysis of the printing paste showing shear thinning; (c–e) SEM images of filaments printed using different nozzle lengths (partial image (a–e) was reprinted from [163]).

insects, where it enhances impact resistance and toughness [164]. Drawing inspiration from nature, alumina-based ceramics were printed in Bouligand configurations by rotating the print direction by 30° for each successive layer. The Bouligand arrangement significantly improves the fracture toughness of additively manufactured ceramics when compared to both transfilament and interfilament configurations (see Fig. 5.6). This enhancement in toughness is primarily due to the increased resistance to crack propagation, as cracks are forced to travel helically rather than following a straight path in Bouligand structures, as illustrated in the figure. The same figure also showcases additively manufactured ceramic gears, highlighting the flexibility and high resolution of DIW technology in producing complex ceramic components.

DIW is ideal for creating porous ceramic structures with periodic features, particularly when high surface quality is not a primary concern. One of the distinctive advantages of this technique is its ability to align micro/nanoscale constituents, enabling the fabrication of mesostructures with unique properties. Despite its advantages, DIW faces challenges such as low resolution and high porosity in the final parts which limit the broader applications of these materials. To address these issues, postprocessing techniques like metal or molten glass infiltration can be applied to enhance the density and improve the structural properties of the fabricated ceramics.

Fig. 5.6: Bioinspired, Bouligand structure. Microstructures of bioinspired platelet-epoxy composites produced via robocasting of platelet pastes and comparison to natural analogues. Schematic diagrams of the three morphologies: (a) transfilament, (b) interfilament, (c) bioinspired, Bouligand structure, (d–f) SEM images of fracture surfaces of the three composites, (g) printed parts to demonstrate the flexibility of the technique, and (h) SEM images the microstructure of the printed part. Scale bars: 10 and 500 μm (image was reprinted from [163]).

5.3 Bulk solid-based technologies

Unlike the powder or liquid-based techniques, bulk solid-based ceramic AM involves the feedstock materials to be in bulk solid form such as a thin layer of ceramic sheet or a filament. Sheet lamination and fused filament fabrication (FFF) are the manufacturing techniques using these solid feedstocks, respectively.

5.3.1 Sheet lamination

Sheet lamination involves using ceramic sheet preforms as the feedstock material. These ceramic sheets are produced through the tape casting method, as shown in Fig. 5.7. In tape casting, ceramic powder is mixed with a solvent, binder, and other additives to create a homogeneous blend. This mixture is then spread onto a surface and thinned using blades, a process commonly referred to as "doctor blading." During the final drying stage, the solvent is evaporated, leaving behind the ceramic film. Tape casting is widely used to fabricate thin sheets not only from ceramics but also from polymers, metals, and paper, with the capability to produce films just a few microns thick.

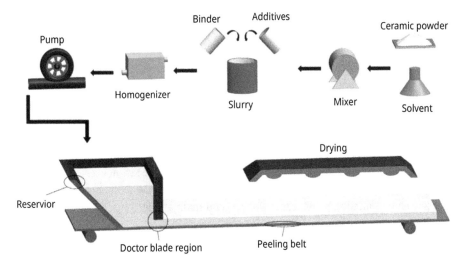

Fig. 5.7: Schematic of tape casting process (figure was reprinted with permission from [165]).

Once the green ceramic sheets are prepared, they are rolled onto the working platform, and any excess material is removed. Subsequent layers are then added until the final 3D part is formed, as detailed in Chapter 1. After the last layer is deposited, the part undergoes binder removal and sintering at elevated temperatures. Various types of ceramics have been successfully fabricated using this technique, including Al_2O_3, zirconia, SiC [166], piezoelectric materials [167], HA for bone implants [168], and ceramic composites like Si/SiC [169] and ZrO_2/Al_2O_3 [170].

Sheet lamination offers advantages such as high speed, versatility with different ceramic materials, and low fabrication temperatures (before sintering). However, it also has some drawbacks, including the need for sheet ceramic feedstock, high porosity, and poor surface adhesion between layers, which can lead to delamination.

5.3.2 Fused filament fabrication (FFF)

The FFF technique involves melting a thermoplastic filament, which solidifies upon deposition onto a print bed, as described in Chapter 1. Since brittle ceramics cannot be shaped into flexible filaments for FFF, ceramic powders are mixed with thermoplastic polymers to create ceramic–polymer composite filaments. These composite filaments, with up to 60% ceramic content, can then be used to fabricate 3D structures using standard FFF printers. A key advantage of this approach is that it allows for the use of low-cost, widely available FFF printers instead of more expensive systems.

After printing, the part undergoes binder removal and sintering to densify the ceramic material. Like other ceramic AM methods, the FFF process is commonly used to create bioceramic scaffolds, taking advantage of the high porosity left by the removed polymer. In addition, ceramics such as alumina, silicon nitride [171], and piezoelectric [172] materials have been successfully printed using the FFF method. Proper dispersion of ceramic particles within the polymer is crucial to achieving high-quality parts with the desired porosity. The size and shape of these particles, along with standard FFF parameters such as nozzle width, layer thickness, build orientation, raster angle, and porosity, all impact the quality of the printed object.

In this chapter of the book, we have explored the AM techniques for ceramic materials under three main categories based on feedstock type: powder-based, slurry-based, and bulk-solid-based ceramic AM. Each method has its own advantages and limitations, which should be considered when choosing the most suitable technology for a particular application. Table 5.1 summarizes these commonly used AM techniques, highlighting their key characteristics, benefits, drawbacks, and main applications.

Tab. 5.1: Comparison of ceramic additive manufacturing technologies (redrawn according to [173]).

Feedstock type	AM method	Resolution	Speed	Surface quality	Feedstock Cost	Applications
Powder-based	Binder jetting	µm–mm	Medium	Medium	Medium	Structural/bio
	Powder bed fusion	µm–mm	Medium	Low	Low	Structural/bio
Slurry-based	Stereolithography	µm	Slow	High	High	Structural/bio
	Digital light processing	µm	Medium	High	High	Structural/bio
	Direct ink writing	µm–mm	Medium	Low	Low	Structural/bio
Solid-based	Sheet lamination	µm–mm	High	Medium	Medium	Structural
	Fused filament fabrication	mm	Medium	Low	Medium	Functional

5.4 Additive manufacturing of polymer-derived ceramics

Polymer-derived ceramics are created by transforming preceramic polymers into ceramics through a heat treatment process. The precursors used in this process are organosilicon polymers, including poly(organocarbosilanes), poly(organosiloxanes), poly(organosilazanes), and poly(organosilylcarbodiimides) [174]. The polymer-to-ceramic conversion occurs by heating these silicon-based polymers to approximately 1,000 °C in an inert atmosphere, causing the C–H bonds to break. This results in the release of volatile compounds such as H_2 and CH_4, which leads to the formation of an inorganic ceramic material [175]. Figure 5.8 demonstrates that different ceramic compounds can be produced by modifying the precursor polymer. For example, SiC is derived from poly(organocarbosilanes), $Si_xC_yO_z$ from poly(organosiloxanes), and $Si_xC_yN_z$ ceramics can be obtained from poly(organosilazanes) or poly(organosilylcarbodiimides).

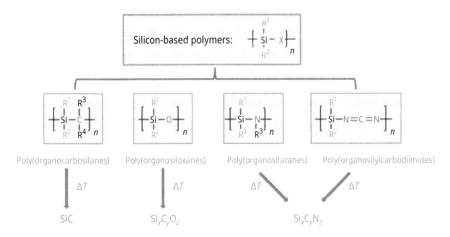

Fig. 5.8: Molecular structures and the types of polymer precursors transforming into ceramic materials (figure was reprinted with permission from [174]).

Polymer-derived ceramics are known for their exceptional properties including high chemical durability, excellent creep resistance, semiconducting behavior, and stability at ultrahigh temperatures (up to 2,000 °C) [174]. As a result, they are widely used in applications that require temperature resistance or as functional materials in micro/nanoelectronics. While polymer-derived ceramics were first discovered over 50 years ago, their popularity has surged in recent years, particularly with advancements in AM technologies. Since the precursor materials are polymers, these preceramic polymers can be additively manufactured into complex geometries and subsequently pyrolyzed to form ceramic parts.

One of the most successful techniques for fabricating these materials is vat polymerization, as photoactive groups like thiol, vinyl, acrylate, methacrylate, or epoxy can be easily attached to organosilazanes [176]. Additionally, metallic precursors can be incorporated into organosiloxane preceramic polymers to produce metal-based ceramics, as demonstrated by recent research [177]. Figure 5.9 illustrates the AM process of polymer-derived ceramic structures, detailing the preparation of photocurable preceramic polymers, DLP printing of these polymers, and the subsequent pyrolysis steps used to fabricate octet truss-shaped polymer-derived ceramics. Figure 5.10 illustrates the designed and fabricated polymer-derived ceramic part in this study, both before and after the pyrolysis steps. As shown, complex lattice structures can be effectively produced using this technology.

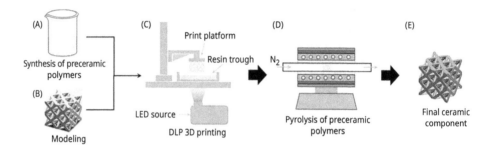

Fig. 5.9: Additive manufacturing process of polymer derived ceramics: (A) precursors used in the printing mix, (b) octet truss structure was designed using CAD software, (c) DLP 3D-printing process, (d) pyrolysis process performed on 3D-printed pre-ceramic polymer, and (e) ceramic material derived after the pyrolysis process (figure was reprinted with permission from [177]).

One of the key advantages of utilizing preceramic polymers in AM is the ability to leverage well-established, high-resolution technologies to fabricate ceramic components. Additively manufactured polymer-derived ceramics, including those doped with metals, also hold promise for innovative applications in areas such as magnetics, catalysis, MEMS, and more. However, a major limitation of AM with polymer-derived ceramics is the significant shrinkage (>20%) that occurs during the polymer-to-ceramic transformation, as depicted in Fig. 5.10.

5.5 Mechanical properties of AM ceramics

As discussed throughout this chapter, high porosity is a characteristic feature of additively manufactured ceramic parts. Depending on the manufacturing method, porosity levels in fabricated components typically range from 20% to 60%. This high porosity significantly compromises the mechanical properties of ceramic parts, limiting their application to areas where porosity is advantageous such as biomedical scaffolding. Due to this elevated porosity, bending strengths of less than 30 MPa are commonly reported for these ceramics, regardless of the material type or manufacturing

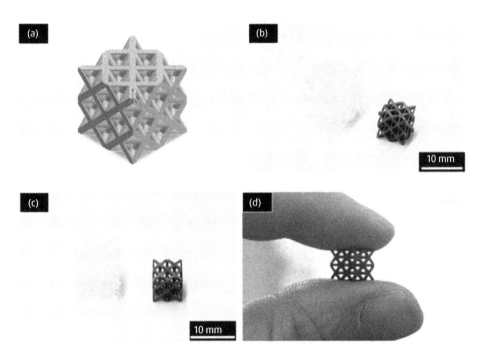

Fig. 5.10: (a) Octet lattice structure designed as CAD model, (b,c) 3D-printed polymer derived ceramic component before and after pyrolysis, and (d) picture of ceramic component showing the shrinkage after pyrolysis at 1,000 °C. (figure was reprinted with permission from [177]).

technique. Figure 5.11 (adapted from Zocca et al. [178]), illustrates the relationship between compressive strength and porosity. The data clearly demonstrate that increasing porosity leads to a substantial reduction in the strength of ceramic materials.

Additively manufactured ceramics intended for structural applications are almost always subjected to postprocessing and densification techniques, such as infiltration, isostatic compaction, and heat treatment, as previously described. These steps enable the production of ceramic components with high density (>90%), significantly enhancing their mechanical properties. Table 5.2 (adapted from Wang et al. [179]) presents the mechanical properties of selected ceramic materials after densification. The data clearly show that the strength of ceramic parts can be dramatically improved through densification, making their mechanical performance comparable to ceramics fabricated by conventional manufacturing methods. However, even after densification, additively manufactured ceramics generally exhibit lower mechanical properties compared to those made via traditional techniques. Fracture strength and toughness are particularly affected, as unavoidable defects and residual porosity persist despite postprocessing.

Ceramic AM continues to be an area of active research due to the inherent advantages of ceramics over metals and polymers including superior thermal and chemical

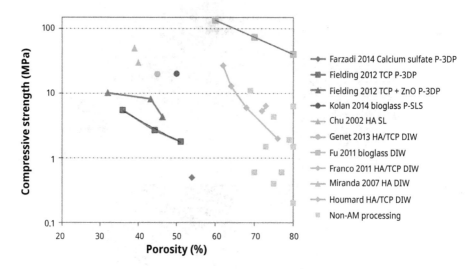

Fig. 5.11: Compressive strength versus porosity of porous ceramics fabricated via additive manufacturing (figure was reprinted with permission from [178]).

resistance, unique mechanical properties, cost efficiency, and biocompatibility. Despite these benefits, additively manufactured ceramic parts have limited applications due to their high porosity, which results from low powder packing density and the formation of voids during the removal of sacrificial organic materials used in the

Tab. 5.2: Comparison of the mechanical properties of additively and conventionally fabricated ceramic materials redrawn according to [179].

Ceramic type	Manufacturing type	Fracture strength (MPa)	Fracture toughness (MPa*m)	References
Aluminum oxide (Al$_2$O$_3$)	Conventional	310–379	4.5	[179]
	Additive	148	–	[180]
	Additive	228	–	[181]
	Additive	–	2.1–4.4	[182]
Zirconum oxide (ZrO$_2$)	Conventional	900	13	[179]
	Additive	731	–	[183]
	Additive	763	6.7	[184]
Silica (SiO$_2$)	Conventional	110–200	0.62–0.67	[179]
	Additive	52.5	–	[185]

Tab. 5.2 (continued)

Ceramic type	Manufacturing type	Fracture strength (MPa)	Fracture toughness (MPa*m)	References
Silicon carbide	Conventional	324	4	[179]
	Additive	160	–	[186]
Boron carbide	Conventional	200	4	[179]
	Additive	58.1–67.4	3.5	[187]
Silicon nitride (Si$_3$N$_4$)	Conventional	679–896	5–8	[179]
	Additive	597	3.97–7.07	[188]
Aluminum nitride (AlN)	Conventional	428	3.5	[179]
	Additive	160	–	[186]

printing process. A straightforward approach to improve density is the use of ceramic slurries instead of dry powders. While slurry-based AM offers advantages over powder-based methods, additional densification steps are still required to produce nearly pore-free parts. Consequently, compared to polymer and metal AM, ceramic AM remains an expensive, time-intensive, and multistep process, limiting its widespread adoption within the AM community.

6 Bioprinting

Bioprinting is the additive manufacturing of living or nonliving biomaterials designed to replicate natural tissues or organs. This innovative field holds great promise for tissue engineering and regenerative medicine, aiming to replace damaged or diseased tissues with artificial alternatives. Additionally, three-dimensional (3D)-printed biomaterials provide valuable platforms for studying biological mechanisms, understanding diseases, and testing drug efficacy on materials that closely mimic in vivo conditions. Consequently, bioprinting finds extensive applications in basic research, biomedicine, and the pharmaceutical industry.

The bioprinting process begins with creating a computer-aided design (CAD) model of the target tissue or organ. This is followed by preparing the bioink, a feedstock material that may consist of natural or synthetic substances tailored to the specific application. The bioink is then deposited layer by layer on a printing platform or an existing biomaterial substrate using computer-controlled algorithms.

To fully grasp the potential and limitations of bioprinting, it is crucial to explore the types of bioink materials currently available, the additive manufacturing techniques employed for their deposition, and the functional outcomes of these biomaterials in in vivo or in vitro settings. This chapter is structured to provide an overview of common bioprinting methodologies, the bioinks used today, and key applications of the technology in biomedicine. The discussion concludes by addressing the challenges and limitations of current bioprinting methods. By reviewing recent advancements in this rapidly evolving field, this chapter aims to inform biomedical researchers about the boundaries of current 3D printing technologies while fostering interdisciplinary collaboration between manufacturing engineers and scientists. This synergy is essential to expand the capabilities and applications of bioprinting, driving innovation in both research and clinical practice.

6.1 Bioprinting methods

Bioprinting technologies are typically classified into four main categories: inkjet printing, extrusion, vat polymerization, and laser-assisted printing. Figure 6.1 provides a schematic representation of these methods, along with a comparison of their key attributes, such as printing speed, cell viability, and investment costs. A brief description of each technique is given below:

Extrusion-based bioprinting: Extrusion is one of the most common bioprinting techniques, where biomaterials are deposited layer by layer through a nozzle using a pneumatic or piston-controlled pressure system (Fig. 6.1A). This process builds a 3D structure by extruding the bioink onto a platform. To ensure the structural integrity of the

https://doi.org/10.1515/9781501520242-006

extruded material, the bioink must possess sufficient viscosity to resist gravitational forces and prevent sagging. Viscosity can be adjusted by introducing curing agents into the bioink, with cross-linking initiated through external stimuli such as heat or light. In some cases, the cross-linking process can occur naturally at room temperature, depending on the bioink's composition and curing reaction.

Extrusion bioprinting is a straightforward technology capable of handling bioinks with high viscosities and high cell densities. It also supports multimaterial printing using multiple nozzles integrated into a single setup. However, the extrusion process can generate shear stress, potentially leading to cell deformation and damage. As a result, cell viability may decrease if process parameters – such as bioink composition, nozzle pressure, and nozzle diameter – are not carefully optimized.

Despite its versatility, the range of materials suitable for extrusion-based bioprinting is currently limited. The bioink must be capable of encapsulating cells, which typically requires the use of hydrogels. These constraints highlight the need for further advancements to broaden the material options available for this method.

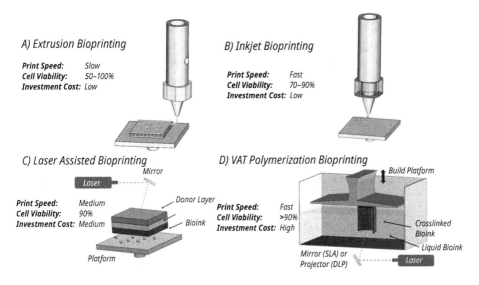

Fig. 6.1: Commonly used bioprinting technologies: (A) extrusion bioprinting, (B) inkjet bioprinting, (C) laser-assisted bioprinting, and (D) vat polymerization bioprinting.

Inkjet bioprinting: This falls under the category of material jetting additive manufacturing, as discussed in Chapter 1. This method involves spraying bioink onto a surface in a drop-on-demand fashion. The deposition of bioink is regulated by either thermal or piezoelectric actuation, as depicted in Fig. 6.1B. In thermal actuation, a heating element near the nozzle increases the temperature of the bioink, creating a bubble. This bubble generates pressure, forcing the bioink out through the nozzle orifice. In piezoelectric actuation, a piezoelectric element produces a pressure pulse that ejects the liquid through the nozzle.

Inkjet bioprinters are relatively inexpensive and straightforward systems, similar to extrusion bioprinters. However, inkjet bioprinting offers faster speeds because multiple printheads can operate in parallel, unlike extrusion printers, which work in series. This parallel operation enables the rapid deposition of multiple cell types.

Inkjet printing delivers high resolution (~30 μm) [189] due to its precise control over drop-wise deposition. Piezoelectric systems also maintain high cell viability, but thermal actuation can damage cells due to the heat applied during printing. A key limitation of inkjet printing is the requirement for low-viscosity bioinks (~0.1 Pa s) [190], which restrict the use of viscous hydrogels and extracellular matrix (ECM) materials. Additionally, cell aggregation and nozzle clogging are frequent issues, especially with high cell-density bioinks. Consequently, inkjet printing is often performed with low cell-density bioinks to ensure consistent printability.

Laser-assisted bioprinting: This is a high-resolution, nozzle-free technique that deposits cells from a source platform onto a substrate platform (Fig. 6.1C). A laser pulse is directed at the bioink, creating a bubble that propels cells onto the substrate. This method enables precise deposition of individual cells, offering resolution superior to other bioprinting methods. Laser-assisted bioprinting is also well-suited for high-viscosity bioinks, as there is no nozzle to clog. The technique achieves excellent cell viability (>95%) since the deposition process avoids shear stress and thermal damage. However, this method has limitations, including a slow printing speed due to the small volume of ink deposited per pulse. Consequently, laser-assisted bioprinting is primarily used for fabricating smaller samples. Additionally, the high cost of laser-assisted bioprinters makes them less accessible compared to other bioprinting technologies.

Vat polymerization bioprinting: This relies on the photopolymerization of bioink composed of light-sensitive polymers. This technique selectively cures the bioink layer by layer to create 3D structures (Fig. 6.1D). Instead of raster scanning a laser line by line (as in stereolithography), light can be projected as a two-dimensional (2D) pattern (as in digital light processing), significantly increasing printing speed compared to other bioprinting methods.

Vat polymerization offers high resolution and dimensional accuracy, making it suitable for fabricating complex biomaterials. While traditional vat polymerization uses UV light due to its high energy, visible light is increasingly favored to prevent DNA damage to cells in the bioink. This method achieves high cell viability (>85%) since the selective cross-linking of bioink via light does not subject cells to shear stress or high temperatures. However, the bioink must be transparent to allow light penetration with minimal scattering. To ensure uniform cross-linking, cell density is often kept low, which can limit the functionality of the printed biomaterial.

Each bioprinting technology offers unique advantages and constraints, making them suitable for specific applications. As bioprinting techniques evolve, further advancements in bioink formulations and process optimization are expected to address current limitations and expand their potential uses in biomedical research and clinical applications.

6.2 Bioink types used in bioprinting

The preparation of bioink is a pivotal stage in the bioprinting process, as its properties – including type, concentration, gelation time, and viscosity – directly influence both printability and the quality of the fabricated structures. An ideal bioink should facilitate the formation of stable 3D constructs with high structural integrity, while supporting cell adhesion, proliferation, and spreading [191]. Additionally, high biodegradability is often desired for applications in tissue engineering and regenerative medicine, making the choice of bioink essential for successful bioprinting.

Hydrogels, biopolymers with tunable properties, are widely used as bioinks because they meet essential bioprinting requirements. These hydrogels can be derived from natural sources, such as collagen, gelatin, alginate, and chitosan, or synthetic sources such as polyethylene glycol (PEG) and polyurethane. Their cross-linking is typically initiated by external stimuli including temperature, pH, light, enzymes, or ions. Cross-linked hydrogels are engineered to be biocompatible, facilitating key cellular activities. The commonly used natural and synthetic bioinks are described in detail below.

Alginate: This is one of the most commonly used bioinks, valued for its low cost and rapid gelation with ionic calcium compounds (e.g., $CaCl_2$, $CaCO_3$, and $CaSO_4$). Its high viscosity improves printability. However, alginate alone does not support cell adhesion and is often mixed with natural polymers like collagen or fibrinogen to enhance this property.

Gelatin: Gelatin, a natural polymer derived from collagen hydrolysis, offers excellent biocompatibility, biodegradability, water solubility, and ease of cross-linking [192]. However, its rapid degradation and poor mechanical strength limit its standalone use. Gelatin methacryloyl (GelMA), a modified gelatin functionalized with methacrylic anhydride, addresses these issues by providing improved mechanical strength and controlled degradation [193]. GelMA requires a photoinitiator for cross-linking, but excessive photoinitiator or UV exposure can harm cells, so careful optimization is necessary [194].

Chitosan: This is a natural polysaccharide biomaterial obtained from the outer skeleton of shellfish or fungal fermentation [195]. It is highly biocompatible, and it has antibacterial properties. Chitosan is widely used in bioprinting of bone, skin, and cartilage tissues. This bioink type, however, suffers from slow gelation time and poor mechanical properties. It can be mixed with gelatin to improve printability and 3D construct shape fidelity [196]. Chitosan hydrogels are widely used in 3D bioprinting, especially for the applications of bone and skin repair/replacement and micro-flow channels fabrication.

Synthetic bioinks: Naturally derived hydrogels have been used extensively as bioinks since they can support cell functions and have biodegradable properties. However, due to their low viscosity, these hydrogels generally possess poor mechanical properties to support tissues, and therefore, synthetic polymers such as PEG is in use for applications requiring high mechanical performance. PEG is a synthetic polymer, and it has tailorable mechanical properties required for bioprinting. It is nontoxic and bioinert. Therefore, cell adhesion is poor in PEG-based hydrogels. Similar to alginate, it is commonly mixed with other active biopolymers to improve cell adhesion properties. Pluronic is another synthetic polymer widely used in bioprinting. Pluronic has similar mechanical and cell adhesion properties to PEG.

ECM-based bioinks: The ECM is a complex network of proteins, glycoproteins, and enzymes that plays a critical role in cellular growth, tissue repair, and remodeling. While traditional bioinks incorporate some of these components, they struggle to replicate the intricate structure of the native ECM. Decellularized ECM (dECM) bioinks are a promising alternative, offering a closer mimicry of natural tissue environments. These bioinks are prepared by removing cells from native ECM using decellularization agents (e.g., detergents, enzymes, or physical treatments) [197]. The resulting dECM can then be mixed with cells to create a bioink that supports cellular growth and differentiation, closely simulating the native microenvironment.

Bioinks vary widely in terms of their mechanical, biocompatibility, and biodegradability properties. They can be composed of natural or synthetic polymers, or even utilize native ECM in decellularized form to better replicate in vivo conditions. Research efforts are actively focused on developing cost-effective bioinks with optimized properties to enhance bioprinting processes and tissue engineering applications. Table 6.1 summarizes the key types of bioinks currently in use, highlighting their primary properties and applications.

Tab. 6.1: Bioink types and properties (redrawn according to [198]).

Ceramic type	Cross-linking mechanism	Bioactiveness	Bioprinting technique
Alginate	Ionic cross-linking with divalent ions and covalent cross-linking with PEG photo-cross-linking with methacrylate	Bioinert	Extrusion, laser-assisted, and inkjet
Collogen	Physical pH- and temperature-mediated cross-linking, chemical cross-linking with genipin/transglutaminase/lysyl oxidase	Bioactive	Laser-assisted and microwave-based

Tab. 6.1 (continued)

Ceramic type	Cross-linking mechanism	Bioactiveness	Bioprinting technique
Gelatin	Physical cross-linking via temperature dependent/opposite charged polymers Chemical cross-linking with horseradish and hydrogen peroxide after modification with phenolic hydroxyl group	Bioactive	Extrusion and laser-assisted
Hyaluronic acid	Chemical cross-linking via coupling of tyramine/NHS/ Cu(I)-catalyzed cycloaddition reaction/host-guest interactions/HA-based copolymer hydrogels/oxidation of thiol-modified HA/thiol-modified HA using PEG derivatives/furan-modified HA derivatives with dimaleimide PEG/gold nanoparticles/photo-cross-linking of tyramine-substituted HA	Bioinert	Extrusion, laser-assisted, microwave-based, and Vat polymerization
PEG	Functionalized through hydroxyl end groups to form PEGDA, PEGDMA, n-PEG photo-cross-linkable hydrogels with different photo-initiators: Irgacure 2959 (365 nm), LAP (visible light), VA086 (405 nm) eosin Y (visible light), and riboflavin with triethanolamine	Bioinert	Vat polymerization, extrusion, and inkjet
GelMA	Chemical photo-cross-linking with different photo-initiators: Irgacure 2959 (365 nm), LAP (405–490 nm), VA-086 (440 nm), Eosin Y (450–550 nm), ruthenium, and sodium persulfate (400–450 nm)	Bioactive	Vat polymerization, extrusion, and inkjet
Pluronic	Temperature-dependent physical cross-linking, and modification of terminal hydroxyl moieties to form photo-cross-linkable hydrogels	Bioinert	Extrusion

6.3 Bioprinting applications

Bioprinting has been utilized across a range of applications, including the fabrication of functional artificial tissues and organs for repair or replacement. It has also been employed to create 3D cancer tissue models, aiding in the study of biophysical mechanisms underlying cancer-related diseases. Key bioprinting applications for functional tissues include the fabrication of vascular structures, blood vessels, neuronal tissues, corneas, cardiac tissues, skin, cartilage, and kidneys, as illustrated in Fig. 6.2. A brief overview of the bioprinting processes for each of these functional tissues is provided below.

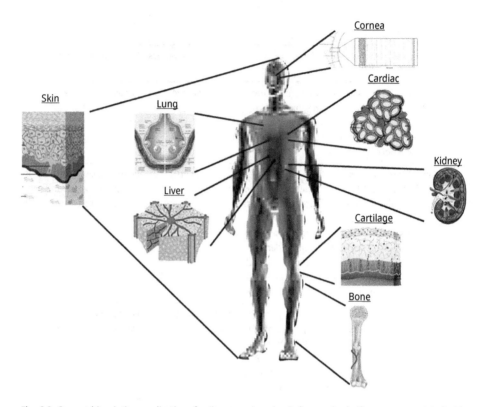

Fig. 6.2: Current bioprinting applications for tissue engineering in human body (figure was reprinted with permission from [198]).

6.3.1 Bioprinting of blood vessels

Blood vessels are critical components of the circulatory system, enabling the transport of oxygen and nutrients to organs and tissues throughout the body via the bloodstream. The design and fabrication of artificial blood vessels have been long-standing goals in tissue engineering. However, success has been largely limited to large arteries due to challenges in replicating small capillaries and mismatches in mechanical properties such as elasticity [199]. Recent advances in bioprinting, including sacrificial, embedded, hollow tube, and microtissue bioprinting techniques, offer promising solutions for creating interconnected microvascular networks with customizable geometries.

Sacrificial bioprinting: This method involves printing a sacrificial hydrogel core surrounded by bioink containing endothelial cells. After printing, the sacrificial material is selectively removed, leaving behind a vascular structure as the surrounding cells form an endothelial layer. While this is the most established method for vascular bioprinting, it is primarily suited for simple geometries and struggles to create complex 3D vascular networks due to the lack of volumetric support.

Embedded bioprinting: In this technique, bioprinting is performed within a viscous hydrogel substrate rather than on a flat print bed. The surrounding hydrogel provides structural support for the bioink as it is extruded, enabling the fabrication of freeform 3D structures. This approach allows for the creation of more intricate and complex vascular networks compared to sacrificial bioprinting.

Hollow tube bioprinting: Hollow tube bioprinting utilizes wet-spinning technology to deposit hollow blood vessel structures in a single step, streamlining the fabrication process. This method simplifies the production of tubular vascular structures, making it efficient and scalable.

The bioprinting of blood vessels is crucial not only for replacing damaged vascular tissues but also for advancing organ bioprinting. Functional organ constructs require an integrated vascular network to transport nutrients and oxygen effectively. Progress in developing bioinks with enhanced biocompatibility and mimicking the properties of native tissues is expected to drive significant advancements in vascular and organ bioprinting, transforming the field of tissue engineering.

6.3.2 Skin bioprinting

The skin is the largest organ in the human body, serving vital functions such as protection, temperature regulation, and sensation. Bioprinting of skin holds tremendous potential for applications in wound healing, replacement of diseased skin tissue, and in vitro testing of medicinal and cosmetic products. The skin's complex and heterogeneous 3D architecture includes structural proteins (e.g., collagen and elastin), various cell types (e.g., keratinocytes and melanocytes), and other specialized structures such as hair follicles, sweat glands, and fatty tissues.

Inkjet bioprinting is one of the most commonly employed techniques for fabricating skin, often utilizing bioinks composed of collagen mixed with skin cells. Significant progress has been made in creating artificial skin with morphological characteristics closely resembling native skin, enabling effective protection and barrier functions [200–203]. Despite these advancements, native skin performs additional essential roles, including temperature regulation (via hair follicles and sweat glands), tactile sensation (through sensory nerves), and skin hydration (via sebaceous glands) [198]. The bioprinting of multifunctional, "smart" skin tissues that replicate these advanced physiological functions is an active area of research, with promising developments underway.

6.3.3 Cartilage printing

Cartilage is a flexible and resilient tissue found at the joints of the skeletal system, where it protects bones by providing cushioning and shock absorption. It also serves as a structural component of the ear and nose. Unlike other connective tissues, cartilage has limited and slow self-repair capabilities due to its avascular nature [204]. Bioprinting technology holds significant potential for cartilage repair and replacement, offering a promising solution to address this limitation.

Chondrocytes, the primary cells responsible for secreting the ECM that maintains and supports native cartilage, are commonly used in bioinks for cartilage bioprinting. However, their limited availability has spurred interest in alternative cell sources, such as mesenchymal stem cells (MSCs). MSCs, which can be isolated from bone marrow, are multipotent cells capable of differentiating into specialized connective tissue cells including those of bone, adipose tissue, and cartilage [205]. Additionally, MSCs exhibit reduced immune rejection, making them particularly suited for bioprinting applications in connective tissue engineering.

A notable example of cartilage bioprinting using MSCs is illustrated in Fig. 6.3 [206]. In this study, human MSCs (hMSCs) were incorporated into a pluronic-alginate bioink mixture. The bioink was printed in the shape of an ear and nose onto a heated platform, where it underwent rapid solidification through the sol–gel transition of Pluronic. The printed structures were subsequently stabilized through alginate crosslinking by immersion in a calcium chloride ($CaCl_2$) solution. This innovative approach highlights the potential of bioprinting to create complex, functional cartilage structures for clinical applications.

6.3.4 Cardiac tissue bioprinting

Heart-related diseases are among the leading causes of mortality worldwide, posing a significant global health challenge. Treating these conditions is particularly complex because the muscular heart tissue (myocardium) in mammals, including humans, has a very limited capacity for regeneration [207]. Currently, heart transplantation is the most effective treatment option, but it comes with substantial limitations such as organ shortages, immune rejection, high costs, and surgical complications. Bioprinting of cardiac tissue offers a promising solution to these challenges, providing potential benefits for patients suffering from heart diseases.

Bioprinting also has transformative implications for drug discovery and development. The ability to create in vitro heart models that mimic the structural and functional properties of the native heart can significantly enhance the evaluation of new treatments. However, bioprinting functional myocardium presents unique challenges, including the need for precise cellular alignment and the integration of contractile capabilities in the printed tissues.

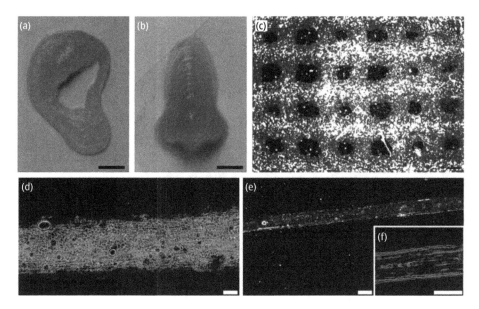

Fig. 6.3: 3D printing of cartilage tissue using alginate hybrid gel: (a, b) post-cross-linking images of a full-sized ear and nose, (scale bars are 1 cm). Wide field fluorescence microscopy of (c) cross-hatch pattern, (d) a single fiber printed through a pipette tip, and (e, f) a single fiber printed through a 30-gauge needle (scale bars are 200 μm; figure was reprinted from [206]).

Recent advances in lab-scale bioprinting have demonstrated promising results in creating cardiac tissues. For example, a study by Noor et al. [208] utilized stem cells derived from a patient, which were differentiated into cardiomyocytes. These cardiomyocytes were incorporated into a hydrogel made from the patient's own dECM. To enhance vascularization, endothelial cells – critical for forming blood vessels – were also included in the bioink mixture. The prepared bioink was printed in the shape of a heart, supported by a sacrificial matrix that provided structural stability during the printing process. After printing, the construct was placed in a bioreactor, where the cardiomyocytes self-organized to form cohesive, functional cardiac tissue. The sacrificial material was then dissolved, leaving behind a fully formed 3D-printed heart model complete with vascular channels capable of carrying blood, as illustrated in Fig. 6.4. This groundbreaking work highlights the potential of bioprinting to create patient-specific cardiac tissues, addressing current limitations in heart transplantation and advancing the development of effective treatments for cardiovascular diseases.

The structure and functionality of bioprinted cardiac patches are assessed through in vitro analyses, with cardiac cell morphology evaluated after transplantation. These analyses reveal elongated cardiomyocytes exhibiting prominent actinin striation. Figure 6.4A illustrates the design of the vascular structure's internal lumen, while Fig. 6.4B and C depicts the embedded bioprinting process for heart tissue. Finally, Fig. 6.4D shows the structure of the bioprinted vascularized cardiac tissue. This

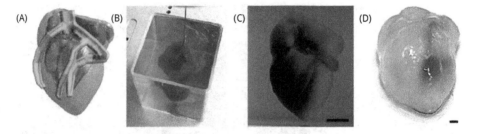

Fig. 6.4: Bioprinting of vascularized heart in small scale: (A) the human heart CAD model; (B, C) a printed heart within a support bath; (D) blue and red dye injection into ventricles after extraction to demonstrate hollow chambers and the septum in-between them (partial figure was reprinted from [208]).

proof-of-concept study, along with other recent research on bioprinting cardiac tissues [209–212], highlights the significant potential of additive manufacturing technologies. These innovations pave the way for personalized tissues and organs as well as more accurate drug screening within anatomically correct and patient-specific biochemical environments.

6.3.5 Kidney tissue bioprinting

Kidneys perform essential functions in the human body including blood filtration, pH regulation, and fluid balance. Similar to the heart, the fabrication of artificial kidneys through bioprinting is a key area of active research, driven by the high demand for kidney transplants and the shortage of available organs. The kidney's complex microarchitecture makes it a suitable candidate for bioprinting, offering the potential for creating functional artificial organs.

3D printing has already been used in clinical kidney transplant surgeries as an educational tool, assisting surgeons in evaluating the feasibility of transplantation and anticipating possible complications. This technology has been particularly useful in the transplantation of adult kidneys to pediatric recipients (<20 kg) with complex structural anomalies [213]. These patient-specific 3D models, fabricated using additive manufacturing, have become crucial in guiding surgeons through clinical kidney procedures.

While current research in functional kidney bioprinting focuses mainly on creating small-scale renal structures, such as the proximal convoluted tubules [198], these efforts are significant for in vitro biotoxicity studies. This is because drug reabsorption and accumulation primarily occur at these sites in the kidney. However, creating a fully functional artificial kidney remains a challenging task due to the organ's intricate microstructure and the need for vascular integration. A recent study [214] investigated the implantation of bioprinted kidney tissue in animals and observed the development of host-derived vascularization, suggesting that the formation of new blood vessels and connection to the host's vascular system is possible for bioprinted kidney tissues in vivo.

6.4 Challenges and limitations of bioprinting functional organs

Bioprinting holds immense potential for creating artificial organs for transplantation. If successful, this technology could revolutionize tissue engineering and offer solutions for a wide range of organ failure-related diseases. However, the bioprinting of functional organs faces numerous limitations and challenges that must be overcome before it can be applied in clinical transplantation.

One of the major hurdles is the lack of vascular connections in bioprinted organs, which are essential for delivering oxygen and nutrients and removing waste via the bloodstream. Current bioprinted organs struggle to form the necessary microvasculature due to their complex 3D structure and small size. While strategies such as in situ printing and methods to connect bioprinted organs with the recipient's existing vascular system show promise, vascularization remains a significant challenge for creating fully functional printed organs.

In addition to vascular tissue formation, bioprinted organs must undergo maturation under physiological conditions before they can be transplanted in vivo. This maturation process, which may take several weeks, requires the use of bioreactors, adding to the cost and slowing down the process. Moreover, ensuring biocompatibility, proper interaction with surrounding tissues, and maintaining the viability and functionality (e.g., contraction, sensing, communication, and filtration) of bioprinted organs in vivo necessitate considerable research investment before the technology is ready for clinical trials.

Given the current limitations, bioprinted artificial tissues and organs are more likely to be used in ex vivo applications such as therapeutic test platforms. These biomaterials, which mimic native tissues and organs, can help researchers better understand disease mechanisms and evaluate the efficacy of novel drugs developed for conditions related to these organs. Additionally, these platforms could reduce reliance on animal models, saving both time and costs in drug testing and disease research. The next section explores one of the major applications of bioprinting in vitro: cancer bioprinting, which could significantly impact cancer treatment and research.

6.5 Bioprinting in cancer research

Despite extensive research efforts, cancer remains one of the most challenging health issues today. The initiation and progression of cancer are heavily influenced by the complex, 3D tumor microenvironment. However, most in vitro tumor models are based on planar, 2D systems, which fail to accurately replicate the conditions found in actual tumors. These simplified 2D models lack the intricate cell and ECM interactions that exist in vivo, which significantly limits the relevance and accuracy of cancer research conducted on them.

To improve the effectiveness of cancer studies, bioprinting offers a promising alternative by enabling the creation of a more realistic 3D tumor microenvironment [215, 216]. This technology allows for controlled distribution of tumor cells within this environment, closely mimicking the in vivo conditions. Previous studies have shown that replacing traditional 2D models with 3D-bioprinted systems can result in significant changes in gene and protein expression, protein gradient profiling, cell signaling, morphology, and cell viability [217–221]. As a result, replicating the 3D cancer microenvironment through bioprinting is crucial for advancing cancer research and understanding tumor initiation and progression. Bioprinting technology has already been applied in various cancer research studies to simulate the cancer microenvironment, offering deeper insights into the fundamental processes behind cancer initiation and growth. A summary of these recent studies utilizing cancer bioprinting, including cell viability data for different methods, is provided in Tab. 6.2.

Tab. 6.2: Recent research studies on cancer bioprinting (redrawn according to [216]).

Cancer type	Bioink	Cell viability	Bioprinting method	References
Breast	Mesenchymal stem cells + gelatin methacrylate	30–50%	Stereolithography	[222]
Breast	hBMSCs + MDA-MB-231/MCF-7	53–80%	Stereolithography	[223]
Breast	MCF-7 + PEG	>90%	Inkjet	[224]
Breast	MDA-MB-231/IMR-90 MCTS	High	Extrusion	[225]
Breast	Breast adenocarcinoma + mouse macrophage + sodium alginate	>90%	Extrusion	[226]
Brain	GSC23 + HMSCs + sodium alginate/gelatin	>90%	Extrusion	[227]
Brain	Human glioma stem cells + gelatin/alginate	87%	Extrusion	[228]
Brain	U118 glioma + pluripotent stem cell-derived neural organoid	N/A	Extrusion	[229]
Brain	GAM + GBM + gelatin methacryloyl/gelatin	High	Extrusion	[230]
Brain	GSC123 + U118 + sodium alginate	>90%	Extrusion	[231]
Cervical	HeLa + gelatin/alginate/fibrinogen	>95%	Extrusion	[232]
Cervical	HeLa + gelatin/alginate/fibrinogen	>90%	Extrusion	[233]
Cervical	HeLa + 10 T1/2 + PEGDA	N/A	Stereolithography	[234]

In these studies, 3D-bioprinting techniques have unlocked numerous opportunities, particularly for in vitro monitoring of cancer progression with enhanced control over the ECM. Bioprinting has greatly improved the biomimicry of cancer models compared to traditional tissue engineering methods, due to its ability to accurately replicate the tumor environment and support improved vascularization [235, 236]. One of the key advantages of additive manufacturing is its ability to create more realistic, precise, and flexible models when compared to conventional 2D techniques.

Recent applications of bioprinting in cancer research have produced significant results, advancing our understanding of cancer biology and opening new avenues for investigating cancer progression, cell interactions, drug efficacy, and treatment strategies. Although bioprinting has been applied to a limited range of cancers, such as breast, brain, cervical, and ovarian (as shown in Table 6.2), expanding its use to other types of tumors and studying the biophysical factors driving cancer initiation and progression could substantially improve drug efficacy and interaction models in future cancer research.

7 Topology optimization

Optimizing the geometry of a structure is crucial for reducing weight, material usage, and manufacturing costs, offering substantial economic and environmental advantages. This is particularly important in aerospace applications, where minimizing weight is critical without compromising functionality. Recent advancements in additive manufacturing have further emphasized the importance of optimization, as this technology eliminates many of the design constraints associated with conventional manufacturing methods. Additive manufacturing allows for the creation of complex geometries beyond the limitations of traditional design spaces.

The geometrical optimization process focuses on identifying the optimal design parameters to achieve objectives such as minimizing mass, volume, and stress while adhering to specific design constraints. Three primary approaches are employed to optimize a design's geometry: size, shape, and topology optimization techniques, as illustrated in Fig. 7.1.

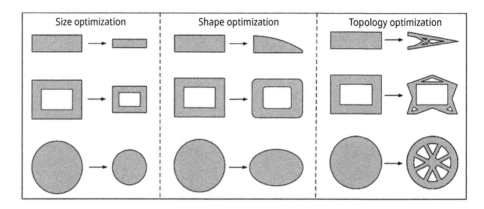

Fig. 7.1: Structural optimization techniques (figure was reprinted from [237]).

In a size optimization process, design parameters such as the length, width, or thickness of components are adjusted iteratively to achieve an optimized configuration. Similarly, shape optimization focuses on modifying the shape of a component to enhance its functionality, such as achieving a uniform stress distribution and minimizing stress concentrations.

Topology optimization, on the other hand, involves determining the optimal connectivity, shape, and placement of internal and external structures within a solid domain by identifying the best material distribution [238]. Unlike size and shape optimization, which work within predefined geometrical variables (e.g., length, thickness, or shape), topology optimization allows for greater design freedom, enabling structures to take on any shape within the design space. This flexibility makes topology optimiza-

https://doi.org/10.1515/9781501520242-007

tion particularly well-suited to complement the manufacturing freedom provided by additive manufacturing. Consequently, topology optimization is widely used in early conceptual and preliminary design phases, where modifications to the topology can significantly enhance the final part's performance.

In practice, these optimization methods are often combined in a multistep process. In the initial phase, topology optimization is employed due to its robustness and flexibility to establish the optimal structural configuration. This is followed by detailed size and shape optimization to refine the design under realistic loading conditions. The accuracy and effectiveness of the initial topology optimization are critical to the success of this combined process.

This chapter delves into the details and significance of topology optimization, presents the primary methods employed in the process, and explains how topology optimization can be integrated into the design workflow using finite element analysis software.

7.1 Topology optimization for additive manufacturing

The mathematical foundation of topology optimization was first applied to simple truss structures in 1904. Over time, it was extended to more complex systems such as beams [239, 240], porous materials, and composite structures [241]. These foundational concepts were later incorporated into numerical algorithms, leading to the development of commercial topology optimization software in the 1990s, which was primarily tailored for subtractive manufacturing methods.

While the mathematical principles of topology optimization have existed for over a century and commercial simulation tools have been available for the past 30 years, the full potential of topology optimization has only been realized with the advent of AM. AM technologies, with their unparalleled design flexibility, have enabled the uncompromised fabrication of highly complex, topologically optimized components.

The general workflow for integrating topology optimization into the design process is illustrated in Fig. 7.2. The process begins with the creation of a computer-aided design (CAD) model of the original design, which is then imported into a topology optimization platform, such as a specialized software or algorithm. A suitable topology optimization method is selected, aiming to refine the geometry while meeting the defined constraints and objectives of the study.

Once the optimization is complete, the resulting domain can be smoothed for improved manufacturability and aesthetics. The smoothed model is then reanalyzed to ensure that it meets the design requirements. If the model fails to meet these requirements, adjustments are made to the optimization algorithm, and the process is repeated until the desired outcome is achieved.

A successful topology optimization process produces designs that outperform the original model. In the final stage, these optimized models can be fabricated using additive manufacturing, leveraging the technology's ability to produce highly complex structures with superior performance characteristics.

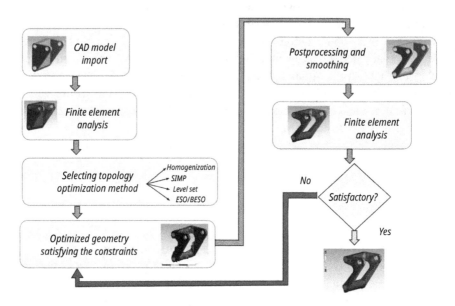

Fig. 7.2: Workflow and processing steps of topology optimization process for additive manufacturing.

Additive manufacturing eliminates the constraints of traditional solid component fabrication, allowing for the creation of highly porous, complex topologies. This capability enables the production of topologically optimized structures with exceptional strength and energy absorption properties at significantly reduced densities. Consequently, mass reduction through topology optimization has found extensive applications, particularly in the automotive and aerospace industries, where weight reduction is a critical objective.

A case study highlighting the use of topology optimization for mass reduction is presented in Section 7.3. In this example, significant weight savings are achieved without compromising the functionality or structural integrity of the designed component. Another critical application of topology optimization in additive manufacturing lies in the design of support structures. Support structures are sacrificial elements required during fabrication to support low-angle or overhanging sections of parts relative to the build plane. While some AM technologies, such as powder bed fusion and binder jetting, do not require support structures, most additive techniques depend on them. Support structures increase printing time and introduce additional postprocessing steps for their removal. According to Liu et al. [242], support structures and their removal can account for 40–70% of the total cost of an AM product. Furthermore, inaccessible support structures in designs with self-contained cavities may remain in place, adding unnecessary weight to the final part. Thus, optimizing support structure design to minimize material use or eliminating the need for supports entirely is crucial for efficient AM.

Figure 7.3 provides examples of optimized support designs aimed at reducing material usage. Figure 7.3A, as reported by Huang et al. [243], compares a traditional straight wall support with a slimmed-down support structure that reduces material consumption and improves printing speed. Figure 7.3B demonstrates a tree-like support pattern that significantly reduces material use compared to solid, straight wall supports.

Overhang-free topology optimization, which eliminates the need for support structures entirely, is an appealing concept for maximizing material efficiency and fabrication speed. However, this approach may compromise the structural performance and functionality of the final part [242]. Additionally, support structures can be advantageous in certain cases, such as enhancing heat transfer to the base plate, which promotes uniform temperature distribution and reduces thermal stresses and warping.

As a practical alternative, strategically placing minimal support material in critical regions can improve part performance without entirely redesigning the geometry to avoid supports. The decision to minimize, eliminate, or combine methods for support structure management depends on the specific design and functionality requirements of the part. Regardless of the approach, topology optimization plays a pivotal role in the design of additively manufactured components where support structures are a factor.

(A)   (B)

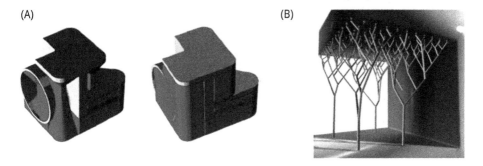

Fig. 7.3: Topology optimized support structures proposed for mass reduction: (A) comparison of the slimmed and straight wall support structures (partial image was reprinted from [243] with permission) and (B) tree-like support structure (partial image was reprinted from [244] with permission).

While topology optimization in additive manufacturing has primarily focused on strength-based weight reduction and support structure design, it has also been extended to other design objectives. One notable application is optimizing additively manufactured thermal systems, such as heat exchangers and heat sinks, utilizing concepts of conductive, convective, and conjugate heat transfer. In these systems, topology optimization modifies geometry-dependent heat transfer coefficients to enhance thermal performance. Additively manufactured thermal components designed through topology optimization demonstrate reduced weight and superior heat transfer capabilities compared to traditional systems [245].

For instance, Lange et al. applied topology optimization to design a heat sink component, aiming to maximize thermal conductivity and minimize weight. The resulting geometry was highly complex and unfeasible for fabrication using conventional manufacturing techniques. However, the design was successfully produced using powder bed fusion additive manufacturing, showcasing the synergistic potential of topology optimization and additive manufacturing. Experimental results from the study confirmed that the topology-optimized heat sink exhibited improved heat transfer performance compared to its unoptimized counterpart.

In summary, topology optimization has been widely applied in additive manufacturing for lightweight designs, reducing or eliminating support structures, and optimizing thermal conductivity to enhance heat transfer and minimize thermal stresses and warping. As additive manufacturing technologies advance and more robust topology optimization methods and numerical simulations become available, additional applications are expected to emerge. The following section discusses the key topology optimization methods currently employed.

7.2 Topology optimization methods

Topology optimization is a mathematical technique used to modify the material distribution within a design space to optimize system performance. Various methods are available, including homogenization, solid isotropic microstructure with penalty (SIMP), evolutionary/bidirectional evolutionary structural optimization, and the level set method. The primary distinction between these methods lies in how the design space is parameterized. Some approaches, such as SIMP, explicitly define the design directly within the finite element domain, while others, like the level set method, represent the design implicitly using a function from which the structure is derived. Despite these differences, all topology optimization methods follow a generalized mathematical framework, which can be described as follows:

$$
\begin{cases}
\text{Objective Function} & \min \text{ or } \max f(x) \\
\text{Constraints} & g(x) \leq 0 \\
\text{Variable Range} & x_{min} \leq x \leq x_{max}
\end{cases}
\tag{7.1}
$$

In this definition, design variable (x) is the vector of independent variables which describes the design. These variables may represent aspects such as geometry, material type, or material distribution.

The objective function $f(x)$ quantifies the performance of the design and returns values indicating its suitability. Common examples of objective functions include weight, displacements in specific directions, stress, or manufacturing costs. In structural optimization problems (e.g., stress-based optimization), the most widely used objective is maximizing structural stiffness or minimizing compliance under a given constraint, such as a specified amount of material removal.

Constraints in the design problem is defined by the function of $g(x) \leq 0$ which may include limits on mass, volume, stresses, displacements, eigenfrequencies, or heat flux. The design space is further defined by the range of the independent variable range (x_{min}, x_{max}). The solution to the optimization problem is typically obtained by solving governing equations expressed in the generic form:

$$K(x)u = F(x) \qquad\qquad (7.2)$$

where $K(x)$ is the stiffness matrix and $F(x)$ is the force vector. Type of the problem determines the governing equations. If the stress-based problem is solved in the design problem, stress–strain constitutive relationships are utilized; in a thermal conductivity design problem, heat equation is used as the governing equation.

Various topology optimization methods are built upon a shared optimization framework, aiming to determine the presence or absence of material within a given design domain. These methods typically adopt the solid-void concept, where "solid" denotes the presence of material and "void" represents its absence. The primary distinction among these methods lies in how the design space is parameterized. Traditional approaches like homogenization and SIMP (solid isotropic material with penalization) define the design explicitly within the finite element domain. In contrast, methods such as the level set method rely on implicit functions to represent the structure's boundaries and domain rather than directly parameterizing the design space. The most commonly used topology optimization methods, SIMP and the level set method, are summarized below:

The SIMP method, a widely adopted density-based topology optimization technique, is known for its simplicity and ease of implementation. As a result, it is the foundation of most commercial topology optimization software [246]. This method determines the optimal material distribution within a design space, accounting for specific load cases, boundary conditions, manufacturing constraints, and performance requirements. In SIMP method, the design variables are defined for each element and normalized value for pseudo-density parameter ranging from 0 (void) and 1 (solid). Density variables are penalized with a basic power law as follows:

$$f(\rho_e) = \rho_e^p\, f_0 \qquad\qquad (7.3)$$

where the objective function, f, is selected as the physical quantities such as material stiffness, cost, or conductivity depending on the problem type. ρ_e is the density of the finite elements used in the optimization and p is a finite penalty parameter commonly used in the range of 1–3. For each element, the assigned relative density can vary between a minimum value ρ_{min} and 1, which allows the assignment of intermediate densities for elements (characterized as porous elements). The penalty factor p diminishes the contribution of elements with intermediate densities (gray elements). The penalty factor steers the optimization solution to elements that are either solid black ($\rho_e = 1$) or void white ($\rho_e = \rho_{min}$). Numerical experiments indicate that a penalty factor value of $p = 3$ is suitable.

The level set topology optimization method is a relatively recent, shape-based optimization technique that uses implicit functions to define structural boundaries, rather than directly parameterizing the design domain. In this method, boundaries are represented by the zero level of the level set function, as shown in Fig. 7.4a and b. The initial shape is defined by the contour where the level set function equals zero, and the optimization algorithm is applied to meet the objective function while satisfying given constraints. During the optimization process, the shape evolves dynamically within the design domain.

The level set optimization produces smoother boundaries compared to the SIMP method. In addition, the level set method does not use intermediate density material (gray zone) in the design domain resulting clear, unambiguous geometries. However, the level set method is a time dependent (as the level set function moves along the domain with a controlled velocity) initial value problem and therefore its accuracy is strongly dependent upon the initial design which is a major drawback of this methodology. Nevertheless, new tools are under development to reduce this dependency [247]. The level set methods also require reinitialization during the process when the level set function is not satisfactory (too flat or too steep) which adds additional computational complexity therefore, algorithms not requiring this process is under development to reduce the computational cost [248].

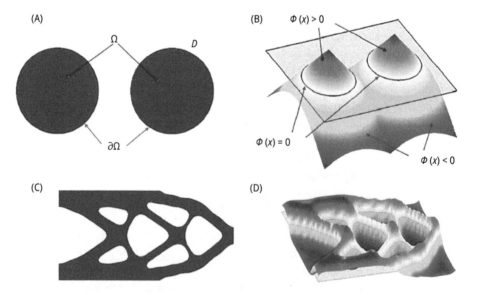

Fig. 7.4: Level set topology optimization method: (A, C) 2D topologies and (B, D) the corresponding level set functions. Image was reprinted from [249] with permission.

7.3 Solution of topology optimization problem using ANSYS finite element software

To demonstrate the integration of topology optimization in additive manufacturing, a bracket design under load is used as an example, as shown in Fig. 7.5A. The bracket is subjected to a 5,000 N force applied at the upper-right pinholes, with the other pinholes serving as cylindrical supports fixed in all directions (radial, tangential, and axial). The objective of this design is to reduce the bracket's mass as much as possible while maintaining its structural stability. The topology optimization process was carried out using ANSYS Workbench finite element analysis software, with the bracket model imported as shown in Fig. 7.5B. A static structural analysis under the specified loading conditions revealed a maximum stress of 100 MPa in the bracket (Fig. 7.5C). The analysis also showed that most of the bracket experienced minimal stress (indicated by the blue zones), demonstrating the potential for topology optimization.

In the initial optimization, the goal was to reduce the mass of the bracket by 40%. The topology-optimized model after a 40% mass reduction is shown in Fig. 7.5D, where material in the blue areas, which did not contribute to structural stability, was removed. The SIMP-based topology optimization algorithm was selected to achieve this by maximizing stiffness (minimizing compliance). After optimization, the design was smoothed for improved manufacturability and aesthetics, as shown in Fig. 7.5E, and reanalyzed using finite element analysis (Fig. 7.5F). The results showed that the maximum stress remained unchanged at 100 MPa despite the 40% reduction in mass.

In the second design iteration, the mass reduction constraint was increased to 75%, resulting in the topology shown in Fig. 7.5G. Similar to the first iteration, the optimized model was smoothed for better manufacturability. The final design, shown in Fig. 7.5H, underwent another finite element analysis (Fig. 7.5I) under the same loading conditions. The analysis revealed a slight increase in the maximum stress to 112 MPa, even after removing a significant amount of mass. The final design was deemed satisfactory, balancing mass reduction, structural stability, and manufacturability.

In conclusion, topology optimization is a powerful tool for designing additively manufactured parts with enhanced performance. This includes benefits such as mass reduction, optimized support structures, and improved mechanical, thermal, and electrical properties. As more accurate and computationally efficient topology optimization methods continue to develop, their integration into the manufacturing of additively fabricated components will be further advanced.

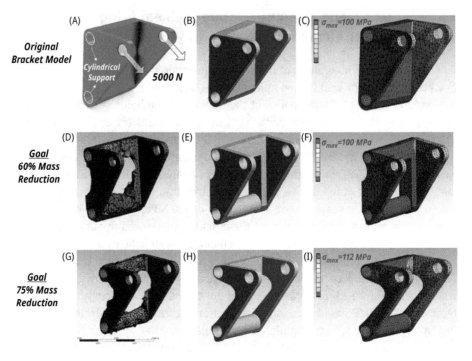

Fig. 7.5: Step-by-step design process for the topology optimized bracket under mechanical loading: (A) loading and boundary conditions for the bracket, (B) imported bracket CAD design into ANSYS finite element software, (C) structural analysis of the original design, (D) 40% mass reduced model obtained by the topology optimization algorithm, (E) smoothed CAD design based on the topology optimized model, (F) structural analysis of the topology optimized design, (G) 75% mass reduced model obtained by the topology optimization algorithm, (H) smoothed CAD design based on the 75% mass reduced model, and (I) structural analysis of the final design.

8 Advanced concepts in additive manufacturing

8.1 Hybrid additive manufacturing

Hybrid additive manufacturing (hybrid-AM) refers to either the multimaterial layer-by-layer manufacturing process or the integration of AM with other manufacturing technologies. This approach enhances the design flexibility inherent in AM by introducing an additional dimension to the manufacturing paradigm. In other words, hybrid-AM combines the strengths of multiple manufacturing processes or materials to improve part performance and functionality. By doing so, it addresses some of the limitations associated with conventional AM methods.

Hybrid-AM processes typically follow a sequential order, rather than being simultaneous, which minimizes adverse interactions and the influence of individual events on one another. This structured approach allows for greater precision and control during the manufacturing process. In this book, hybrid-AM technologies are explored across three major application fields: additive/subtractive hybrid-AM, hybrid-AM of multimaterial electronics components, and hybrid-AM in tissue engineering. The benefits and limitations of these hybrid methods are discussed in detail in the following sections.

8.1.1 Additive/subtractive hybrid manufacturing

As discussed throughout this book, AM has significantly enhanced the design flexibility of traditional subtractive manufacturing methods. However, subtractive manufacturing retains distinct advantages such as superior surface finish and higher production speed. These complementary strengths have led to the development of hybrid manufacturing systems that integrate additive and subtractive techniques.

Computer numerical control (CNC)-based subtractive manufacturing is particularly well-suited for integration with AM due to their shared reliance on digital control platforms. This compatibility enables seamless coordination between the two processes. Figure 8.1 illustrates the concept of hybrid-AM/subtractive manufacturing, where CNC milling is applied to the surfaces of an additively manufactured part. This combination improves surface quality, enhances precision, and significantly increases processing speed [250].

In addition to milling, other subtractive manufacturing techniques such as turning and drilling can be integrated with AM, further expanding the potential of hybrid-AM systems. This hybrid approach offers distinct advantages over traditional AM by combining the design flexibility of AM with the superior surface finish achieved through subtractive processes.

https://doi.org/10.1515/9781501520242-008

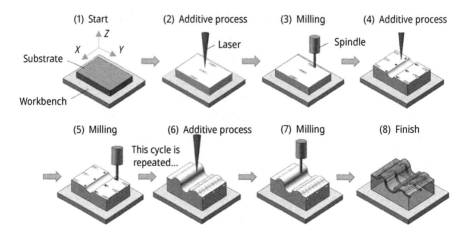

Fig. 8.1: Schematic of additive/subtractive hybrid manufacturing (image was reprinted from [250] with permission).

One significant benefit of hybrid-AM is the elimination of the stair-step effect, a common geometrical artifact in additively manufactured parts. By machining these surfaces during the subtractive phase, hybrid-AM can produce smooth and precise geometries. Additionally, hybrid-AM is particularly advantageous for working with rare or hard-to-machine materials. These materials can first be fabricated to near-net shape via AM, reducing waste and lowering production costs, especially for high-cost materials. As discussed in Chapter 4, the dynamic strength of metals is heavily influenced by surface roughness. Hybrid-AM enables the fabrication of metallic components with enhanced dynamic strength by significantly improving surface quality, making it a superior choice for critical applications [251].

8.1.2 Additive/additive hybrid manufacturing

The integration of subtractive and AM technologies aims to enable robust, high-speed production of metal parts with complex geometries and superior surface finishes. Similarly, combining different AM techniques offers a complementary approach to leveraging the strengths of each technology within a unified manufacturing system. Given the unique benefits and limitations of various AM methodologies, an effective hybrid approach integrates high-resolution, single-material AM processes with low-resolution, multimaterial capabilities. For example, researchers at the University of Texas at El Paso [252] developed a direct write (DW) integrated hybrid manufacturing system that allowed the printing of electrical circuitry onto structures fabricated using stereolithography (SLA). This concept was later expanded by integrating fused filament fabrication (FFF) to build 3D circuit boards for a CubeSat satellite [253]. Additionally, material jetting was combined with FFF to fabricate electronic structures at high resolutions

(~10 µm) [254]. These developments illustrate the potential of hybrid additive/additive systems that integrate multimaterial techniques, such as direct ink writing and material jetting, with robust, high-resolution methods such as FFF and SLA.

The primary applications of hybrid-AM/AM technologies are in the electronics industry. These systems often require the integration of metal conductors, ceramic insulators, dielectric materials, polymers, and semiconductors to create multimaterial electronic components with reduced cost and weight. High geometrical complexity, demanded in sectors such as automotive, medical, and aerospace, poses challenges to the production of traditional printed circuit board (PCB)-based electronics. An example of a hybrid process combining SLA and direct ink writing is demonstrated in a study by Jo et al. [255], which details the fabrication of a functional 3D-printed PCB device with complex geometry. As shown in Fig. 8.2, the circuit diagram of the electronic system (Fig. 8.2A) begins with SLA fabricating an exterior wall by photopolymerizing liquid polymer material around electronic components. The process is paused to clean the partially printed part and dispense interconnects using direct ink writing. During this step, electronic components such as batteries, resistors, and diodes are also placed on the printed surfaces. SLA then resumes to embed these components within the polymer, completing the structure (Fig. 8.2B–D).

This integration of multimaterial additive processes into high-resolution AM systems demonstrates how hybrid-AM technologies address the growing demand for complex, functional electronics in advanced applications.

Hybrid-AM/AM technologies for manufacturing electronic components hold immense potential due to their exceptional design flexibility and ability to fabricate multimaterial systems with minimal material waste. However, a key limitation of these methods lies in the intermittent nature of the process between different manufacturing steps, which significantly reduces production speed. Advancing hybrid-AM technologies to enable simultaneous integration of multimaterial AM techniques will greatly enhance their efficiency and accelerate their adoption for a wide range of innovative applications in the future.

8.1.3 Hybrid additive manufacturing/scaffolding technologies

AM plays a crucial role in tissue engineering, enabling the creation of patient-specific 3D implant structures and scaffolds. However, the absence of surface textures that mimic native tissues negatively impacts cell adhesion and proliferation, limiting the application of AM for certain tissue engineering needs [257]. Additionally, the low resolution of current AM technologies hinders the fabrication of submicrometer structures resembling the natural extracellular matrix (ECM) and hierarchical porous architectures with multimodal pore size distributions [258]. Hybrid-AM technologies, which integrate standard AM processes with cell scaffolding techniques, offer a solution by optimizing the surface morphology of printed structures for improved cell ad-

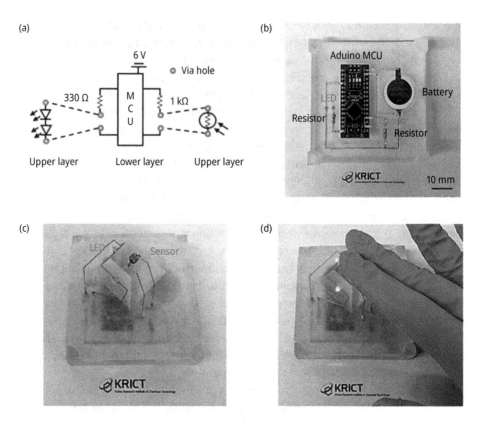

Fig. 8.2: Additively fabricated printed circuit board (PCB) system: (a) circuit diagram of a 3D-printed; (b) bottom view of the PCB with an Arduino-programmed MCU, a photosensor, a battery, resistors, via-holes, and LEDs, (c) top view of the printed PCB, (d) image of the 3D-PCB operation, showing that the LED turns on when the photosensor is covered (figure was reprinted with permission from [256]).

hesion and proliferation. Furthermore, multimaterial, multiscale (micro-nano-macro) materials can be deposited onto printed structures, enhancing the biophysical properties of these 3D-printed biomaterials. Techniques like electrospinning and freeze-drying are particularly suitable for integration with AM, providing additional functionality and precision to these hybrid systems.

Freeze-drying methods can be seamlessly integrated with AM technologies to create structures featuring multiscale (micro- and nano-level) pores. These custom 3D architectures with hierarchical pore arrangements provide an expanded surface area, enhancing cell adhesion and proliferation [259]. Freeze-drying is a well-established process involving material freezing, pressure reduction, and solvent removal via sublimation. This method can be effectively combined with DW AM to produce 3D tissue scaffolds. In this hybrid-AM approach, a polymer fiber solution is dispensed at low temperatures using DW techniques, and the solvent is subsequently freeze-dried, leaving behind a porous fiber network. The scaffold's porosity can be precisely controlled by adjusting process parameters such as temperature. Natural biopolymers, like collagen, can be utilized in this process to create hierarchical 3D scaffolds with highly porous surfaces, providing an ideal environment for cell entrapment and fostering tissue growth.

Wet-spinning offers an alternative to freeze-drying for fabricating porous fiber networks in 3D. This technique involves extruding a polymer solution into a coagulation bath, where the filament solidifies and is collected in predefined layer-by-layer patterns [260, 261]. The resulting fibers exhibit "spongy" morphologies, which enhance their biological response, making them well-suited for tissue engineering applications. One key advantage of this hybrid manufacturing approach is its simplicity, as it achieves the desired porous structures in a single step, unlike the more complex freeze-drying-based processes. However, its applicability is limited to specific biomaterials, which restricts its versatility across broader applications.

Electrospinning is a widely utilized polymer fiber production technique that employs electric forces to draw charged polymer solutions into fibers at nano- and microscale dimensions. When integrated with AM, electrospinning can form hybrid-AM technologies in various configurations. For example, electrospun fibers can be deposited onto sacrificial targets fabricated via AM, enabling the creation of micropatterned scaffolds with random fibers and defined 3D surface microtopography. Alternatively, melt electrospinning offers a direct fiber deposition method similar to FFF. This process involves melting thermoplastic polymers during electrospinning rather than dissolving them in a solvent. Direct deposition eliminates the need for sacrificial targets, enabling precise, single-step fiber placement on the print bed with high resolution. Hochleitner et al. [262] recently demonstrated this approach, producing uniform thermoplastic fibers in the submicron range using melt electrospinning-based hybrid-AM. However, this method requires a sophisticated setup, including integrated heating and electrical insulation, which increases complexity and cost. Moreover, the high temperatures involved can negatively impact heat-sensitive materials such as collagen or growth factors.

To address these challenges, electrohydrodynamic jet printing (E-jetting) has emerged as a promising alternative. This technique uses ethanol as a target material to produce highly porous 3D scaffolds with controlled filament orientation at room temperature [263]. Studies have shown that fibrous scaffolds fabricated through E-jetting, consisting of entangled micro- and nanofibers, exhibit superior cell seeding efficiency and adhesion compared to scaffolds produced by extrusion-based AM [257].

In summary, integrating fiber scaffolding techniques with AM methods like FFF or direct writing results in advanced scaffold structures ideal for cell adhesion and proliferation in tissue engineering. These 3D fiber-based constructs closely mimic the native ECM in both topography and spatial arrangement. By adjusting the scaffolding technique and process parameters, attributes such as porosity, fiber size, and orientation can be precisely controlled. Table 8.1 summarizes the key hybrid-AM scaffold fabrication techniques, highlighting their advantages and limitations.

Tab. 8.1: Summary of hybrid-AM technologies used in tissue engineering (redrawn according to [257]).

Hybrid-AM method	Advantages	Limitations	Materials	Applications
Freeze drying	High porosity with smaller pore size Control of porosity level	Two-step process	PLGA Chitosan PLLA/TCP Type-I collagen PLLA/chitosan	Bone Cartilage Nerve
Wet spinning	Spongy fiber morphology One-step process	Limited set of biomaterials	PCL PCL/HA Star poly(ε-caprolactone)/ HA	Bone
Melt electrospinning	Submicron fiber size and micron size fiber-to-fiber distance Microscale threads consisting randomly interwoven micro/ nanofibers	Complex experimental setup (heating + electrical insulation)	PCL	Skin Neural Vascular TE
E-jetting	Different strut morphology depending on the viscosity of target solution	Difficult control of micro/ nanosized strut size	PCL	Bone

8.2 Additive manufacturing of thermoelectric materials

The thermoelectric effect is a fascinating phenomenon that involves the direct conversion of heat energy into electricity, offering significant potential for addressing waste energy recovery challenges. A substantial portion of the energy generated in various processes is ultimately lost as waste heat, emphasizing the importance of efficient recovery systems. Thermoelectricity refers to the ability to convert thermal energy directly into electrical energy (and vice versa) without relying on moving parts or working fluids. Thermoelectric materials, which generate electricity when exposed to a temperature differential, are particularly promising for reducing waste heat. When a thermal gradient is applied – such as engine heat on one side and ambient temperature on the other – these materials produce an electric potential, ΔV, described by the following equation:

$$\Delta V = S \Delta T \qquad (8.1)$$

where S and ΔT are Seebeck coefficient and relative temperature, respectively. Heavily doped semiconductors are the most effective thermoelectric materials due to their combination of high Seebeck coefficients, high electrical conductivities, and low thermal conductivities. Low thermal conductivity is particularly important to maintain the temperature gradient across the material and minimize energy loss. Thermoelectric materials can be classified as either p-type or n-type based on their doping. When exposed to a thermal gradient, p-type materials exhibit the movement of positive charges (holes) from the hot side to the cold side, while n-type materials facilitate the movement of negative charges (electrons) in the same direction. When these materials are electrically connected in series and thermally connected in parallel, an electric current is generated, converting the temperature gradient into electrical energy, as illustrated in Fig. 8.3A. While a single thermoelectric module produces only a limited amount of electrical power, higher power outputs can be achieved by arranging multiple modules in series, as shown in Fig. 8.3B. Applications with high power demands, such as NASA's radioisotope thermoelectric generators used in various space missions, require thousands of modules to generate electrical power in the kilowatt range.

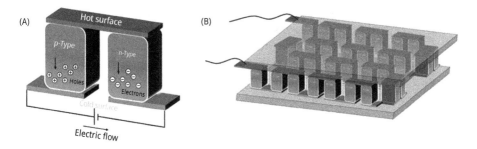

Fig. 8.3: Principle of thermoelectric energy generation: (A) thermoelectric effect in p- and n-type thermoelectric materials and (B) assembled modules in a thermoelectric generator device.

Since its discovery over a century ago, thermoelectricity has attracted significant scientific interest and has been extensively studied by researchers worldwide. Despite this, its applications have largely been confined to niche areas such as NASA's deep space exploration systems described earlier. The limited use of thermoelectric systems in everyday life is primarily due to two fundamental challenges: low energy conversion efficiency and the complex, multistep fabrication process required even for simple geometries. The conventional fabrication of thermoelectric devices involves multiple steps including powder synthesis, ingot preparation, dicing, metallization, leg formation, and final assembly, as illustrated in Fig. 8.4. This intricate process contributes to the difficulty of scaling thermoelectric technologies for broader applications [264].

The fabrication and assembly of hundreds or even thousands of modules to produce a single thermoelectric device presents significant challenges in conventional manufacturing processes. Additionally, traditional manufacturing methods are limited to flat thermoelectric module geometries, making it difficult to fabricate complex leg configurations such as butterfly, trapezoid, and cross-shape with enhanced thermoelec-

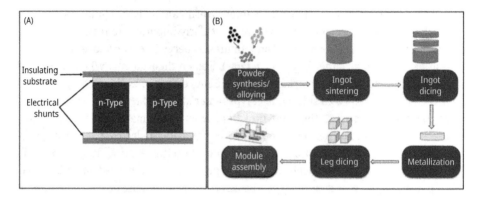

Fig. 8.4: (A) Schematic of thermoelectric unicouple system with *n*- and *t*-type legs and (B) processing steps for the fabrication of a conventional thermoelectric device (image was reprinted from [264] with permission).

tric conversion efficiency [265]. Similarly, recently developed cascade [266] and annular [267, 268] thermoelectric systems also offer higher efficiencies and AM holds transformative potential to fabricate these intricate configurations and enable the production of next-generation thermoelectric systems with higher efficiency, reliability, and reduced cost.

AM technologies, such as extrusion, vat polymerization, and powder bed fusion, have recently been applied to the fabrication of thermoelectric materials. Beyond providing flexibility in device size and geometry, AM can significantly reduce the cost and time required for thermoelectric system production. This is largely because the same AM platform can simultaneously fabricate electrical contacts and deposit thermoelectric materials, allowing for a fully automated manufacturing process. Several notable advancements in this field have been developed recently.

For instance, in 2015, He and colleagues [269] utilized SLA to fabricate thermoelectric structures. They mixed *p*-type bismuth antimony telluride (Bi0.5Sb1.5Te3, BST) powder with a photocurable resin in a vat and used SLA to 3D print components by curing the photopolymer embedded with thermoelectric powder. The fabrication steps are shown in Fig. 8.5. Their custom setup enabled the preparation of components with a high thermoelectric powder loading of up to 60% by weight. Postprinting, the components were thermally annealed at 350 °C for up to 6 h, enhancing their thermoelectric properties. The efficiency of thermoelectric materials is measured by the thermoelectric figure of merit (*ZT*), with higher *ZT* values indicating better conversion efficiency. In this study, a maximum *ZT* value of 0.12 was achieved for samples fabricated using SLA and annealed for 6 h, demonstrating the potential of AM in advancing thermoelectric material fabrication.

In 2019, Oztan et al. [270] demonstrated the fabrication of bismuth telluride (Bi_2Te_3) in complex geometries using FFF, a technique that enables AM of thermoelectric materials with the help of a sacrificial polymer matrix. The process begins by mixing acrylo-

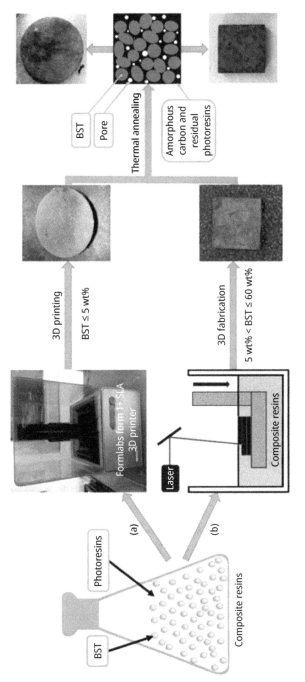

Fig. 8.5: Stereolithography additive manufacturing of thermoelectric materials (image was reprinted from [269] with permission).

nitrile butadiene styrene (ABS) with Bi_2Te_3 powders and extruding the mixture into a thermoelectric filament precursor, as illustrated in Fig. 8.6. This filament feedstock can then be 3D-printed into custom geometries using a standard desktop FFF printer.

Similar to the SLA process, postprocessing heat treatment is essential to optimize the thermoelectric performance of the FFF-printed samples. During this postprocessing, the sacrificial ABS polymer in the composite is first removed. The remaining Bi_2Te_3 material is then sintered below its melting temperature (585 °C) to enhance its properties. In this study, the fabricated samples achieved a thermoelectric ZT of 0.54 at room temperature, a significant improvement compared to the SLA method. The higher efficiency observed in FFF may be attributed to several factors, including the greater concentration of thermoelectric material in the printed composites (80% vs. 60% by weight) and the heat treatment process conducted at elevated temperatures (up to 550 °C), which promotes stronger bonding between particles and improves electrical conductivity and thermoelectric performance. However, as with SLA-fabricated samples, high porosity was noted in the FFF-produced components. Minimizing this porosity remains critical to further enhancing thermoelectric conversion efficiency.

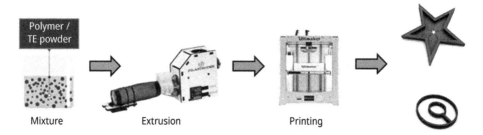

Fig. 8.6: Schematic of FFF additive manufacturing of thermoelectric materials [270].

DW (paste extrusion) AM technique was also utilized to fabricate thermoelectric materials in viscous ink form in 2018 [271]. Unlike conventional approaches that utilize organic printing aids, this study introduced Sb_2Te_3 chalcogenidometallate ions as inorganic binders to prepare printable Bi_2Te_3-based inks. The use of inorganic binders resulted in thermoelectric inks with pronounced shear-thinning behavior, enabling successful paste extrusion as detailed in Chapter 1. After extrusion, the printed specimens underwent heat treatment to promote bonding between thermoelectric powders and enhance conversion efficiency. The resulting materials exhibited uniform thermoelectric properties, with dimensionless ZT values of 0.9 for p-type and 0.6 for n-type materials, comparable to their bulk counterparts. Despite its advantages, paste extrusion for thermoelectric material fabrication has certain limitations. These include the necessity of postprocessing heat treatment and significant shrinkage (~20%) during the heat treatment process, which can impact the final dimensions and structural integrity of the printed components.

The challenges of high porosity, low efficiency, postprocessing requirements, and filament fabrication in thermoelectric material production can be effectively addressed using powder bed fusion technologies, such as selective laser sintering (SLS) and selective laser melting (SLM). These laser-based techniques offer promising solutions for the direct fabrication of thermoelectric materials with enhanced performance and efficiency. In 2015, Leblanc et al. introduced the concept of laser-based powder bed fusion for thermoelectric materials, demonstrating the feasibility of rapid prototyping of thermoelectric compounds using SLM technology [272]. By optimizing process parameters, such as laser power, they achieved a ZT of 0.11 in a subsequent study. The schematic of the SLM system used by the Leblanc group, along with the fabricated Bi_2Te_3 components, is shown in Fig. 8.7.

Further advancements in 2017 utilized SLM to fabricate fine (~3.4 μm) n-type $Bi_2Te_{2.7}Se_{0.3}$ powders, achieving a maximum ZT of 0.84 [273]. This study demonstrated that laser-based powder bed fusion can produce thermoelectric materials with conversion efficiencies comparable to those fabricated using conventional methods.

Qiu et al. [274] fabricated $Bi_{0.4}Sb_{1.6}Te_3$ bulk materials using slurry-based SLM technology. This process begins with the preparation of thermoelectric powder through thermal explosion followed by ball milling. Part of the powder is spark plasma sintered to create a bulk substrate material for laser sintering, while the rest is mixed with alcohol to form a thin slurry layer (~50 μm). The alcohol is evaporated by heating, leaving behind a powder layer that is then laser sintered. This method achieved a high ZT of 1.1 at 316 K, coupled with impressive mechanical strength of 91 MPa. The enhanced performance is attributed to reduced grain size and a high density of dislocation defects caused by rapid solidification, which significantly strengthened the material's mechanical properties. This work highlights the capability of SLM technologies to produce Bi_2Te_3-based thermoelectric materials with high texture, robust mechanical properties, and excellent thermoelectric performance. It also provides valuable insights for applying these techniques to other layered material systems.

SLS has also been explored for the AM of thermoelectric components, offering a distinct approach compared to SLM. Unlike SLM, SLS partially melts the material, as described in Chapter 1, resulting in the formation of porous structures. Shi et al. [276] recently utilized SLS to fabricate porous thermoelectric samples of $Bi_{0.5}Sb_{1.5}Te_3$ (BST). The partial melting of powders during the SLS process inherently produced porous thermoelectric structures. Remarkably, the SLS-fabricated BST exhibited a high ZT of 1.29 at 54 °C. This superior performance is attributed to the significantly reduced thermal conductivity resulting from the material's porosity, along with the boundaries and defects introduced during the SLS process. These features enhance the thermoelectric efficiency compared to bulk BST materials fabricated using conventional methods.

Bismuth telluride-based materials have been a primary focus in thermoelectric studies due to their exceptional performance at room temperature, cost-effectiveness, and extensive prior research. However, tellurium's rarity and high toxicity present

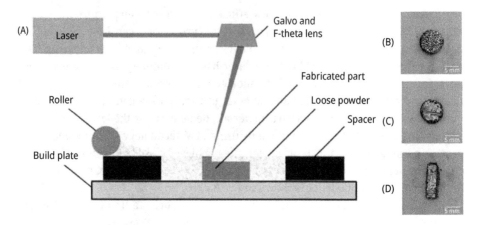

Fig. 8.7: (A) Schematic of SLM additive manufacturing procedure of thermoelectric materials, (B) disk shape printed Bi_2Te_3 part with 16 W laser power (relative density ~81%), () disk shape-printed Bi_2Te_3 part with 25 W of laser power (relative density ~88%), and (D) rectangular shape printed Bi_2Te_3 part (figure was reprinted with permission from [275]).

significant challenges including elevated manufacturing costs and safety concerns. Consequently, research efforts are increasingly directed toward discovering alternative thermoelectric materials to overcome these limitations and improve conversion efficiency. Notable advancements include studies on PEDOT:PSS, Half-Heusler alloys, copper selenide (Cu_2Se), and copper sulfide (Cu_2S).

Inkjet-printed PEDOT:PSS films on paper demonstrated a relatively low ZT of 0.013 [277]. Yan et al. investigated the thermoelectric properties of ZrNiSn-based Half-Heusler alloys using powder bed fusion (SLM) and achieved a ZT of 0.39 [278]. Copper-based thermoelectric materials have garnered significant interest due to their abundance, nontoxicity, and usability at mid-range service temperatures (300–600 °C). For instance, Choo et al. [279] prepared Cu_2Se-based viscous inks and achieved a ZT of 1.2 using 3D DW paste extrusion. Similarly, Gustinvil et al. [280] used paste extrusion to print Cu_2S-based thermoelectric inks, achieving a ZT of 1. However, the diffusion of copper at elevated temperatures remains a challenge, potentially impacting the reliability of these materials in high-temperature applications.

In summary, AM shows tremendous potential as an alternative method for fabricating thermoelectric devices. It simplifies the manufacturing process, minimizes material waste, and enables greater geometric flexibility compared to conventional approaches. Table 8.2 summarizes key studies employing AM for thermoelectric material production. While bismuth telluride has dominated initial research, exploration into AM-compatible materials with improved performance, reduced costs, lower toxicity, and greater earth abundance is steadily expanding. Continued innovation in thermoelectric materials, manufacturing techniques, and postprocessing technologies is essential for advancing thermoelectric systems and broadening their applications in energy conversion.

Tab. 8.2: The comparison of research studies reported on additive manufacturing of thermoelectric materials.

AM method	Printer type	Material	Maximum ZT	References
SLA	Custom and commercial (Form1) 3D printer	$Bi_{0.5}Sb_{1.5}Te_3$	0.12	[269]
FFF	Commercial 3D printer (Ultimaker2)	Bi_2Te_3	0.54	[270]
Paste extrusion	Custom 3D printer	$Bi_{0.4}Sb_{1.5}Te_3$ (p-type)	0.9 (p-type) 0.6 (n-type)	[271]
SLM	Custom 3D printer	Bi_2Te_3	0.11	[275]
SLM	Custom 3D printer	$Bi_2Te_{2.7}Se_{0.3}$	0.84	[273]
Slurry-based SLM	Custom 3D printer	$Bi_{0.4}Sb_{1.6}Te_3$	1.1	[274]
SLS	Commercial 3D printer (XJRP SLS400)	$Bi_{0.5}Sb_{1.5}Te_3$	1.29	[276]
Material jetting	Commercial 3D printer (DIMATIX DMP-381)	PEDOT:PSS	0.013	[277]
SLM	Custom 3D printer	ZrNiSn	0.39	[278]
Paste extrusion	Custom 3D printer	Cu_2Se	1.2	[279]
Paste extrusion	Custom 3D printer	Cu_2S	1	[280]

8.3 4D printing with smart materials

The concept of 4D printing was first introduced by Tibbits during a TED talk in 2013 [281]. It is defined as an evolution of 3D printing, incorporating a time variable that enables a 3D-printed structure to change its shape, properties, or functionality over time [282, 283]. Essentially, 4D printing extends traditional 3D printing by integrating an additional dimension (time) into the process. Like 3D printing, 4D printing involves key steps such as CAD modeling, layer-by-layer fabrication, and postprocessing. However, what sets 4D printing apart is the application of external stimuli to induce predictable, time-dependent transformations in the 3D-printed structure. This process relies on the use of smart materials that respond to specific external stimuli, such as heat, light, electricity, moisture, pH, or magnetic fields, as illustrated in Fig. 8.8. These stimuli trigger the smart materials to transition into another stable state in a controlled and predictable manner.

A crucial aspect of 4D printing is not merely achieving time-dependent changes but doing so in a precisely predictable way. This requires extensive mathematical modeling to anticipate the material's behavior under specific stimuli. The resulting transformation depends on the type of smart material, the configuration of the 3D-printed structure, and the external stimulus applied. By carefully selecting and combining these elements, 4D printing enables structures to shift between states in a controlled, predictable fashion, paving the way for innovative applications in various fields.

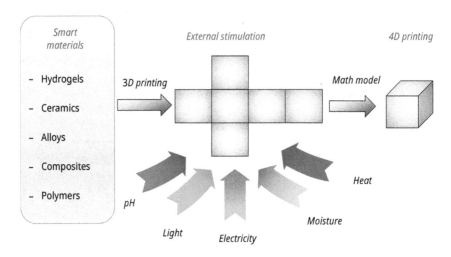

Fig. 8.8: Schematic representation of the 4D printing procedure including the smart material and external stimulation configurations.

External stimuli are essential to activate changes in smart materials, enabling them to alter their shapes, properties, or functionalities. Common stimuli used in 4D printing include heat (temperature), moisture (water), light, electricity, magnetic forces, and pH. In some cases, multiple stimuli are applied simultaneously (such as a combination of heat and water) to achieve specific transformations or configurations. The choice of stimulus is determined by the intended application, which also dictates the type of smart material integrated into the 4D-printed structure.

As illustrated in Fig. 8.8, the materials utilized in 4D printing are broadly categorized similarly to those in 3D printing: hydrogels, ceramics, alloys, polymers, and composites. However, these materials are able to respond to external stimuli which classifies them as "smart" materials. Material selection is a critical step in the 4D printing process, as it directly influences the functionality of the printed structure. Smart materials used in 4D printing exhibit unique capabilities, such as self-sensing, decision-making, responsiveness, shape memory, self-adaptability, multifunctionality, and self-repair. The major types of smart materials employed in 4D printing, along with their specific properties and applications, are detailed in the following section.

8.3.1 4D printing materials

As mentioned earlier, 4D printing materials possess the unique ability to alter their properties in response to external stimuli. Although the range of materials available for 4D printing is more limited compared to those used in traditional 3D printing, recent advancements have significantly expanded the repertoire of stimuli-responsive smart materials. The primary materials utilized in 4D printing applications include hydrogels, shape memory polymers (SMPs), liquid crystal elastomers (LCEs), and active composites.

8.3.1.1 4D-printed hydrogels

Smart hydrogels utilized in 4D printing leverage the interaction between inactive rigid polymers and active soft polymers within a printed system. When an external stimulus is applied to this bilayer configuration, the soft hydrogel component swells or contracts in response, while the rigid polymer maintains its shape. This interplay induces bending deformation within the structure. By arranging these materials in a predesigned sequence, more complex motions, such as folding and unfolding, can be achieved.

Hydrogel-based smart materials typically respond to two primary stimuli: moisture (water immersion) and temperature (heat). Hydrophilic hydrogels swell upon water exposure, and these rubber-like materials can induce significant transformations in 3D-printed bilayer structures. Figure 8.9A showcases an example of the shape-changing capabilities of water-responsive, 4D-printed hydrogel structures, as demonstrated in a pioneering study by Tibbits [283]. In this work, hydrogel-based materials were configured as a series of hinges that enabled a flat linear structure to transform into a 3D cube when immersed in water.

Similarly, water-responsive hydrogel materials have been employed to fabricate 4D-printed parts mimicking plant cell wall structures [284], as illustrated in Fig. 8.9B. This design featured a bilayer composition of a soft acrylamide hydrogel (active layer) combined with rigid cellulose fibrils (stiff layer). The composite structures were produced using a DW extrusion technique, with bilayers oriented at specific angles relative to the petal's long axis (e.g., 90°/0° and −45°/45°). Immersion in water caused the hydrophilic material to swell, resulting in the intricate geometrical transformations depicted in the figure.

Temperature-responsive hydrogels have also been developed as smart material feedstock for 4D printing. Poly(*N*-isopropylacrylamide) (PNIPAm) is a widely used thermoresponsive material for such applications. In aqueous solutions, PNIPAm hydrogels exhibit hydrophilic behavior and swell at temperatures below 32 °C. However, when the temperature exceeds this threshold, the hydrogel becomes hydrophobic, leading to dehydration, shrinkage, and shape transformation. Similar to water-responsive hydrogel composites, temperature-responsive hydrogels can be 3D-printed

 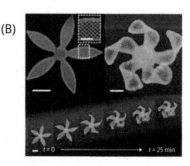

Fig. 8.9: Examples of 4D printing of hydrogel materials: (A) line to cube transformation of a water-immersed hydrogel structure (partial image was reprinted from [283] with permission) and (B) biomimetic 4D printing of a flower with multiple petals (partial image was reprinted from [284] with permission).

alongside nonresponsive polymers to create bi-layered composite hinge systems. For instance, Wu and coworkers fabricated temperature-responsive hydrogels with disproportional swelling properties [285]. These composite hydrogels demonstrated bending deformation in aqueous environments at temperatures above 32 °C due to differential swelling behaviors of their components. Additionally, 4D-printed composite hydrogels have been shown to enable reversible shape deformations. Naficy et al. [286] reported the use of PNIPAm-based hydrogels that responded to both hydration and temperature, showcasing the versatility of these smart materials in achieving dynamic transformations.

8.3.1.2 Shape memory polymers

SMPs have garnered significant interest in 4D printing due to their unique capability to transform their shape under external stimuli and revert to a permanent shape once the stimulus is removed. Unlike hydrogels, SMPs exhibit gradual and controlled transformations, making them ideal for applications requiring morphing structures [287]. For instance, an example of shape change using PCL macromethacrylate SMPs through temperature variation is depicted in Fig. 8.10A [288]. Eiffel Tower and bird figurines were 3D-printed using the SLA technique in a heated vat at 90 °C. As the temperature of the printed structures changed, they displayed reversible shape transformations under thermal stimuli.

Multiple SMPs or combinations of SMPs with nonresponsive materials can be 3D-printed in precisely designed configurations, including specific shapes, placements, and mixing ratios, to create complex geometries. When SMPs with varying properties are integrated into a multi-SMP system, they exhibit selective and distinct shape changes in response to external stimuli, enabling precise control. A notable category of these systems is digital SMPs, which employ SMPs with different glass transition temperatures to achieve temperature-dependent shape changes. Figure 8.10B illustrates an example where a USPS mailbox was designed using digital SMPs that responded to thermal stimuli [289]. In this application, digital SMPs were combined with thermally nonresponsive materials to achieve the desired configurations. Moreover, multistep programming of digital SMPs allows for incremental temperature adjustments, resulting in multiple distinct configurations of the printed structures.

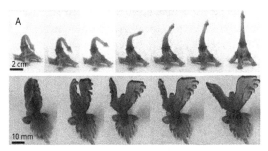

Fig. 8.10: (A) Thermally activated single material shape memory polymers. Eiffel tower and bird figurines respond to temperature stimulus changing their shapes (partial image was reprinted from [288] with permission); (B) time lapse of the folding process of a mailbox shaped digital SMPs (image was reprinted from [289]).

Although thermally responsive SMPs are the most common type, there are also other SMPs that respond to various stimuli, including pH changes, solvents, and moisture (chemoresponsive materials), electrical signals (electroresponsive materials), stress or pressure (mechanoresponsive materials), magnetic fields (magnetoresponsive materials), and light (photoresponsive materials) [290].

8.3.1.3 Elastomer actuators

SMPs are commonly preferred for 4D printing due to their high elastic deformation, low cost, low density, and potential biocompatibility. However, most SMPs exhibit a one-way response, requiring additional programming steps for further active responses. Hydrogels, on the other hand, offer reversible shape changes through swelling and shrinking when stimuli are applied and reversed. Despite this advantage, hydrogels have slower response times, making their shape transformation processes relatively slow. LCEs are unique materials that provide rapid and reversible shape changes, making them increasingly popular as 4D-printed smart materials. Combining the stretchability of elastomers with the self-assembly properties of liquid crystals, LCEs are ideal for fabricating smart stretchable structures, such as soft robots, implantable biomedical devices, and systems based on artificial muscles. LCEs undergo shape changes by transitioning between the liquid crystal (nematic) state and the isotropic state in response to stimuli such as light, heat, or electrical and magnetic fields.

By cycling above and below their nematic-to-isotropic transition temperature (TNI), LCEs alternate between their aligned (anisotropic) and nonaligned (isotropic) states, causing significant and reversible shape changes [291]. Recent work by Kotikian et al. [292] extensively investigated the reversibility of 4D-printed LCEs under thermal stimuli. Figure 8.11 from this study illustrates the transformation process from 2D to 3D, where 3D-printed LCE polymers in flat 2D geometries transition into 3D cone (Fig. 8.11A) and saddle (Fig. 8.11B) configurations as temperatures rise above TNI. These transformations achieved extraordinarily high out-of-plane deformations (~1,628%). Similarly, Fig. 8.11C and D demonstrate isotropic shrinkage and 3D conical array expansion of LCEs under thermal stimuli, respectively. This study also showed that upon cooling, these 4D-printed LCE materials exhibited reversible shape morphing, underscoring their versatility and potential for various applications.

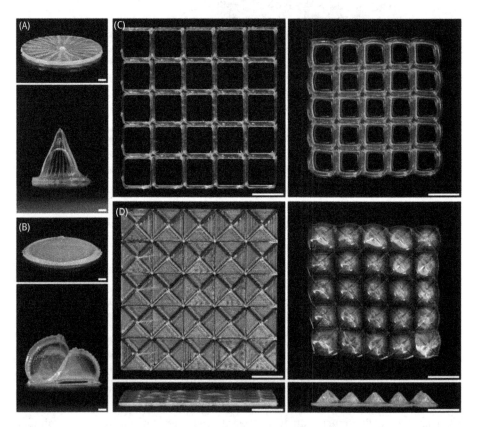

Fig. 8.11: Shape morphing in LCEs under external stimuli. 2D to 3D transformation of disk-shaped LCEs (≈0.4 mm thick) into (A) cone and (B) saddle configurations, (C) top-down images of mesh-shaped LCEAs (≈0.5 mm thick) after printing (left) and shrinking into an isotropic form (right) upon heating above TNI (scale bars = 5 mm), and (D) top and side views of a LCEA sheet after printing (left) and morphing (right) into a conical array upon heating above TNI (scale bars = 5 mm) (figure was reprinted with permission from [292]).

8.3.2 Applications of 4D-printed structures

Predicting the precise state of a material in response to applied stimuli enables unique properties in 4D-printed structures such as self-assembly, multifunctionality, and self-repair. These capabilities offer significant advantages including reduced volume and enhanced transportation efficiency. Smart structures can be printed in compact forms, such as 2D sheets or compressed configurations, and subsequently transformed into larger 3D shapes as needed. This approach minimizes packaging requirements and transportation challenges.

4D-printed structures have promising applications across various fields, including biomedical, electronics, and robotics. In the biomedical domain, devices such as stents and adaptive scaffolds can be fabricated using 4D printing. These systems can be pre-shaped into temporary forms, inserted into the body via minimally invasive procedures, and then deployed into their functional configurations through externally induced stimuli. Another noteworthy application is targeted drug delivery, where drugs encapsulated within 3D-printed structures are released only upon reaching specific locations, triggered by interactions with the targeted tissue or organ.

In aerospace applications, self-assembling 4D-printed components are particularly beneficial. These smart structures can remain folded during transport and expand into their functional configurations, such as antennas or satellite components, when needed. Similarly, self-assembling construction systems could be employed to build structures in remote locations, such as war zones or outer space, with minimal human intervention [293].

4D-printed structures also hold unique potential in soft robotics. For example, Fig. 8.12 illustrates a gripping mechanism achieved using a 4D-printed structure, as reported by Ge et al. [294]. These actuators require no wiring or mechanical hinges, enhancing their simplicity and reliability. Consequently, they are cost-effective and, when

Fig. 8.12: Snapshots of a 4D-printed SMP gripping a bolt. Image was reprinted from [294].

precisely controlled, suitable for delicate interactions with fragile objects [295]. This versatility underscores the transformative potential of 4D-printed materials and systems across diverse applications.

Self-repair is another remarkable property explored in 4D-printed structures. These materials can reorganize their internal structure in response to external stimuli, allowing them to repair damaged areas and restore functionality. Several studies have demonstrated the ability of materials used in 4D printing to self-heal, offering promising applications in areas such as repairing damaged pipes in plumbing systems and creating self-healing vascular models for tissue engineering.

This chapter provides an overview of the concept of 4D printing, the types of materials used, and key applications. Despite its extraordinary potential, 4D printing remains in the development stage since its inception in 2013. Continued research is essential to refine the technology and unlock its full potential for revolutionary applications.

8.4 Artificial intelligence in additive manufacturing

Artificial intelligence (AI) refers to the capability of machines to perform tasks that usually require human intelligence, such as speech recognition, natural language understanding, and decision-making. A key subset of AI, known as machine learning (ML), uses algorithms that allow computers or machines to learn from data and enhance their performance over time. This section of the book focuses on the implementation of ML-based AI applications in AM and examines the reasons behind their rapid growth and increasing adoption.

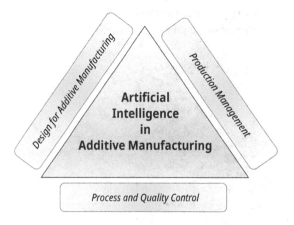

Fig. 8.13: Implementation of AI on AM applications.

AM, as a digital manufacturing technology, generates substantial amounts of data during planning, design, production, and postprocessing stages. Consequently, AI technologies have already had a significant impact on the AM industry. Numerous AI applications have been documented in both academic literature and commercial manufacturing platforms. In this book, we categorize AI applications in AM into three primary areas, as depicted in Fig. 8.13: design optimization, process control, and production management. While other applications may exist outside this classification, the majority of current implementations align with these three categories.

8.4.1 Implementation of AI in design optimization

As emphasized throughout this book, particularly in Chapter 1, Design for Additive Manufacturing (DfAM) fundamentally diverges from traditional design principles by offering unprecedented design freedom. DfAM leverages 3D printing technologies to produce intricate, lightweight components that consume less material than conventional manufacturing methods. Additionally, it enables faster, more cost-effective, and efficient production processes. AI integration into AM further enhance DfAM by optimizing key manufacturing design parameters such as topology, material microstructure, build orientation, support structures, and single print line geometry.

AI-driven DfAM facilitates the optimization of material topology, improving structural, thermal, and electrical performance. Researchers employ AI techniques, including artificial neural networks and deep learning, to refine geometries and dimensions, enabling high-performance topologies under specific loading conditions. For instance, Fig. 8.14 demonstrates a design optimization example. As detailed in Chapter 7, standard topology optimization methods, such as SIMP and level set, are used to maximize stiffness for bearing external loads. However, in applications like bone scaffolds, maximizing overall stiffness may hinder bone regeneration and lead to strain shielding issues [296]. To address this, targeted stiffness designs have been proposed, matching scaffold stiffness to host bone properties. Wu et al. [296] utilized ML techniques to design nonuniform lattice structures (Fig. 8.14) that achieved more uniform and targeted strain fields compared to uniform lattice topologies. These optimized designs closely mimic natural bone responses under complex anatomical and physiological conditions, fostering a biomechanical environment conducive to effective bone formation.

AI is also transforming composite material development for AM. Researchers have employed AI to optimize fiber orientations in fiber-reinforced composites, maximizing their mechanical and thermal performance. Similarly, bioinspired materials designed using ML have been additively manufactured by Gu et al. [297]. As illustrated in Fig. 8.15, a family of unit cells was initially created composing of materials with high (stiff) and low (soft) elastic moduli. By strategically arranging these unit cells, AI-optimized material responses to specific mechanical loading conditions. The resulting 3D-printed materials demonstrated enhanced strength and toughness compared to individual material properties, showcasing the potential of ML-driven material design for AM.

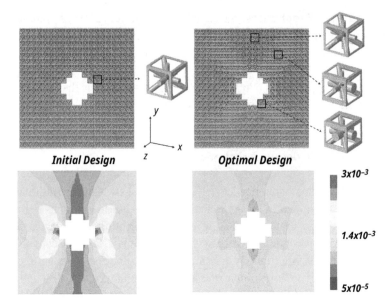

Fig. 8.14: Comparison of the optimized and the initial designs of the squared plate with a center hole. Contour plots represent the strain field under compression (image was reprinted from [296]).

Additionally, ML methods have been applied to minimize shape deviations in additively manufactured parts, addressing critical challenges in improving product quality. Geometrical inaccuracies, including warping, shrinking, and shape changes, can be mitigated by ML, which adjusts the initial 3D model to compensate for deviations. Neural network-based ML algorithms have been used to establish relationships between process parameters and geometry-related errors [298, 299], while Gaussian process learning has achieved over 90% prediction accuracy in capturing global and local shape deviations [300]. These advancements reduce the need for postprocessing, significantly lowering manufacturing costs and time.

In summary, the integration of machine learning into DfAM is revolutionizing AM by advancing design optimization and addressing longstanding challenges. ML methods expand the design space, optimize manufacturing parameters, and improve component performance and printability. These innovations demonstrate AI's transformative potential in material development and part fabrication, unlocking possibilities unattainable through conventional approaches.

8.4.2 Implementation of AI in process/quality control

AI-enabled systems leverage sensors and cameras to monitor the 3D printing process, detecting and correcting anomalies during fabrication. This approach is widely ap-

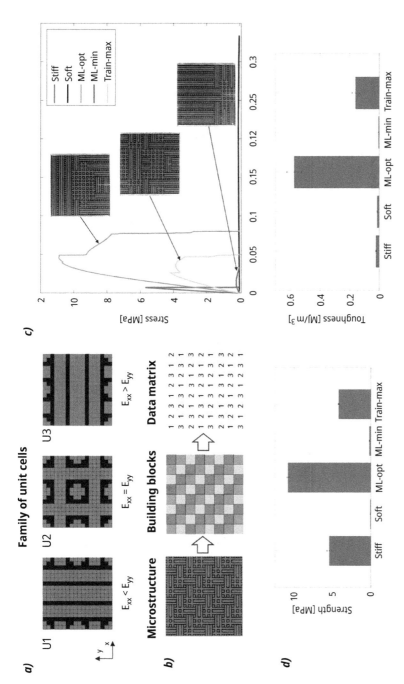

Fig. 8.15: AI-driven material design for AM. (A) a family of three unit cells are considered with variable anisotropic properties. Pink and black colors refer to stiff and soft building blocks, respectively. (B) the microstructure consists of a detailed assemblage of unit cells that is then converted to a data matrix of building blocks encoding the individual unit cells (blue = U1, orange = U2, yellow = U3). (C, D) Mechanical test results of the optimized materials for AM. Reproduced from Ref. [297] with permission from the Royal Society of Chemistry.

plied in AM, particularly in metal powder bed fusion processes, where ensuring part quality and minimizing defect density are of paramount importance. These systems can capture detailed information at each printing layer such as temperature distribution, melt pool morphology, and porosity. This layer-by-layer analysis allows for real-time assessment of part quality. Components with defects or anomalies can be identified and discarded immediately after production, significantly improving reliability, and reducing quality control costs. An example of AI implementation of process monitoring is shown in Fig. 8.16. In this study, Nguyen et al. [301] used powder bed fusion to print stainless steel cubes and used a semisupervised ML approach to detect anomalies occurred during printing in these specimens. These anomalies (lack of fusion, overheating, unfused powder) were also correlated to postprocess characteristics including surface roughness, morphology, or tensile strength in these printed parts.

Fig. 8.16: Representative predictions of the ML model on cubes printed with varying printing parameter settings (image was reprinted from [301]).

Beyond passive quality monitoring, AI can also actively intervene the manufacturing process and minimize the defect formation. By dynamically adjusting printing parameters such as laser power, feed rate, or temperature, AI ensures consistent part quality throughout production. This capability is particularly valuable for maintaining uniformity in mass manufacturing. AI-integrated process control systems are increasingly adopted in commercial powder bed fusion systems, driving interest in their potential to produce reliable components at scale.

8.4.3 Implementation of AI in additive manufacturing automation and process management

One of the primary challenges in AM today is its relatively low processing speed compared to traditional manufacturing techniques. However, this limitation can be mitigated by eliminating labor-intensive steps and automating the entire manufacturing process. Thanks to the inherently digital nature of AM, automation can be seamlessly integrated, with AI playing a critical role in process management and automation.

As an example of automation with AI, Wright et al. [302] demonstrated the use of convolutional neural network-based ML algorithms to determine optimal printing process parameters, such as print speed, line width, and layer height, and fully automated the DW printing process. This eliminated the need for manual process parameter optimization. As shown in Fig. 8.17, the team successfully fabricated composite parts, including the University of Miami logo and a palm tree, using a fully AI-controlled DW AM process. This approach ensured high printing quality and superior mechanical performance.

Postprocessing operations are equally critical in the AM workflow. These include support removal, heat treatment, surface finishing, and quality inspection. Advances in automation, such as ML-guided robotic systems, have significantly streamlined postprocessing. Tasks like support removal and surface polishing can now be performed with minimal human intervention, enhancing both efficiency and consistency in part quality. This highlights the value of automation in reducing manufacturing time and ensuring repeatability.

As the demand for 3D-printed components grows, especially in high-stakes industries like aerospace and medical devices, integrating AM into broader manufacturing workflows is becoming increasingly important. AI-powered systems enhance production planning and resource allocation by managing multiple 3D printers, overseeing material inventories, and scheduling post-processing operations, resulting in smooth and efficient production flows.

In conclusion, AI is revolutionizing AM by improving production efficiency, optimizing designs, and ensuring quality assurance. Its transformative potential spans the entire manufacturing pipeline, from part design to machine operation and quality management. As AI continues to evolve, it is poised to play a pivotal role in shaping the future of AM, driving innovation and productivity to new heights.

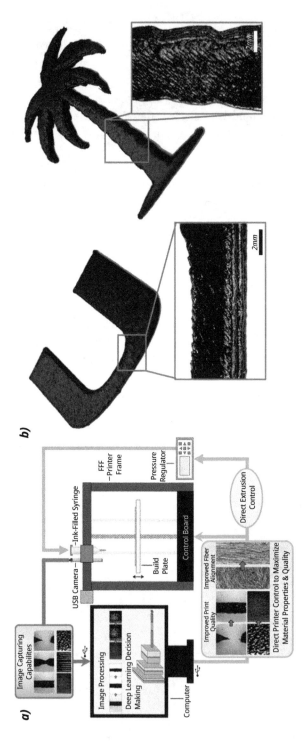

Fig. 8.17: (A) ML-integrated DW AM process; (B) 3D-printed University of Miami logo and palm tree created via an AI-driven autonomous DW process (image reprinted from [302]).

References

[1] Pan, Y., et al., *Taxonomies for reasoning about cyber-physical attacks in IoT-based manufacturing systems*. International Journal of Interactive Multimedia and Artificial Intelligence, 2017. **4**(3): pp. 45–54.

[2] Lim, S., et al., *Developments in construction-scale additive manufacturing processes*. Automation in Construction, 2012. **21**: pp. 262–268.

[3] Chambon, P., et al., *Development of a range-extended electric vehicle powertrain for an integrated energy systems research printed utility vehicle*. Applied Energy, 2017. **191**: pp. 99–110.

[4] Yap, Y.L. and W.Y. Yeong, *Additive manufacture of fashion and jewellery products: a mini review*. Virtual and Physical Prototyping, 2014. **9**(3): pp. 195–201.

[5] (ASTM), A.S.f.T.a.M., *Standard Terminology for Additive Manufacturing Technologies*, in *ASTM F42 – Additive Manufacturing*. ASTM International 2012: Conshohocken, PA 1-3.

[6] Bourell, D., et al., *Materials for additive manufacturing*. CIRP Annals-Manufacturing Technology, 2017. **66**(2): pp. 659–681.

[7] Mendes-Felipe, C., et al., *State-of-the-art and future challenges of UV curable polymer-based smart materials for printing technologies*. Advanced Materials Technologies, 2019. **4**(3) 1800618.

[8] Tumbleston, J.R., et al., *Continuous liquid interface production of 3D objects*. Science, 2015. **347**(6228): pp. 1349–1352.

[9] Shusteff, M., et al., *One-step volumetric additive manufacturing of complex polymer structures*. Science Advances, 2017. **3**(12) eaao5496.

[10] Sher, D. https://www.3dprintingmedia.network/what-is-volumetric-3d-printing-and-why-it-could-mean-the-end-of-additive-manufacturing/?. 2018.

[11] Waheed, S., et al., *3D printed microfluidic devices: enablers and barriers*. Lab on a Chip, 2016. **16**(11): pp. 1993–2013.

[12] Zenou, M. and L. Grainger, *3 – Additive Manufacturing of Metallic Materials*, in *Additive Manufacturing*, J. Zhang and Y.-G. Jung, Editors. 2018, Butterworth-Heinemann. Amsterdam, Netherlands pp. 53–103.

[13] ISO/ASTM 52900:2021 Additive manufacturing — General principles — Fundamentals and vocabulary", International Organization for Standardization (ISO) 2021, ASTM International 2021

[14] Pierson, H.A., et al., *Mechanical properties of printed epoxy-carbon fiber composites*. Experimental Mechanics, 2019. **59**: 843–857.

[15] Compton, B.G., et al., *Electrical and mechanical properties of 3D-printed graphene-reinforced epoxy*. JOM, 2018. **70**(3): pp. 292–297.

[16] Compton, B.G. and J.A. Lewis, *3D-printing of lightweight cellular composites*. Advanced Materials, 2014. **26**(34): pp. 5930–+.

[17] Ahn, B.Y., et al., *Planar and three-dimensional printing of conductive inks*. Journal of Visualized Experiments, 2011(58) e3189.

[18] https://engineeringproductdesign.com/knowledge-base/powder-bed-fusion/. 2019.

[19] Langnau, L. https://www.engineering.com/which-is-faster-for-3d-printing-directed-energy-deposition-or-powder-bed-fusion/. (2018).

[20] Norfolk, M. and H. Johnson, *Solid-State Additive Manufacturing for Heat Exchangers*. JOM, 2015. **67**(3): pp. 655–659.

[21] O., S.M.a.N. *DMLS – Development History and State of the Art*, in *Laser Assisted Net Shape Engineering Conference (LANE 2004)*. 2004. Erlangen, Germany.

[22] Silbernagel, C. *Additive Manufacturing 101–4: What is material jetting?* 2018.

[23] Kvernvik, M. *3D printing used to save two-year-old's life with kidney transplant*. 2018.

[24] Boland T., W.W., Xu T., *Ink-jet printing of viable cells*. 2003: USA, Patent No: US7051654B2

[25] Jones, R., et al., *RepRap – the replicating rapid prototyper*. Robotica, 2011. **29**: pp. 177–191.

https://doi.org/10.1515/9781501520242-009

[26] Quick, D. *The Urbee Hybrid: The World's First 3D Printed Car*. 2010.

[27] V., C. *3D Printed Meat, Is the Future of Meat Meatless?* 2019; Available from: https://www.3dnatives.com/en/3d-printed-meat-040620194/.

[28] Wie, J.J., *Photomechanical materials, composites, and systems: wireless transduction of light into work. Photomechanical effects in amorphous and semicrystalline polymers*, ed. T. J. White. 2017: Wiley Hoboken, USA.

[29] Wudy, K. and D. Drummer, *Aging effects of polyamide 12 in selective laser sintering: Molecular weight distribution and thermal properties*. Additive Manufacturing, 2019. **25**: pp. 1–9.

[30] Schmid, M., A. Amado, and K. Wegener, *Polymer Powders for Selective Laser Sintering (SLS)*. Proceedings of Pps-30: The 30th International Conference of the Polymer Processing Society, 2015. **1664**.

[31] https://www.simplify3d.com/support/materials-guide/properties-table/. 2019.

[32] https://www.3dxtech.com/. 2019.

[33] Lu, S.X., P. Cebe, and M. Capel, *Thermal stability and thermal expansion studies of PEEK and related polyimides*. Polymer, 1996. **37**(14): pp. 2999–3009.

[34] Singh, S., C. Prakash, and S. Ramakrishna, *3D printing of polyether-ether-ketone for biomedical applications*. European Polymer Journal, 2019. **114**: pp. 234–248.

[35] www.polymerdatabase.com. 2015.

[36] https://www.msesupplies.com/pages/list-of-thermal-expansion-coefficients-cte-for-natural-and-engineered-materials. 2019.

[37] www.dupont.com. 2019.

[38] Liu, F.Y., et al., *Polyimide film with low thermal expansion and high transparency by self-enhancement of polyimide/SiC nanofibers net*. RSC Advances, 2018. **8**(34): pp. 19034–19040.

[39] Mcconnell VP. Resins for the hot zone, part 1: polyimides. High Perform Compos. 2009; 7: 31–42

[40] Chandrasekaran, S., et al., *3D printing of high performance cyanate ester thermoset polymers*. Journal of Materials Chemistry A, 2018. **6**(3): pp. 853–858.

[41] Rahim, T.N.A.T., et al., *Comparison of mechanical properties for polyamide 12 composite-based biomaterials fabricated by fused filament fabrication and injection molding*. Translational Craniofacial Conference 2016, (TCC 2016), 2016. **1791**.

[42] Rodriguez-Panes, A., J. Claver, and A.M. Camacho, *The Influence of Manufacturing Parameters on the Mechanical Behaviour of PLA and ABS Pieces Manufactured by FDM: A Comparative Analysis*. Materials, 2018. **11**(8): 1333.

[43] Cantrell, J.T., et al., *Experimental characterization of the mechanical properties of 3D-printed ABS and polycarbonate parts*. Rapid Prototyping Journal, 2017. **23**(4): pp. 811–824.

[44] Mansour, S., A. Gilbert, and R. Hague, *A study of the impact of short-term ageing on the mechanical properties of a stereolithography resin*. Materials Science and Engineering A – Structural Materials Properties Microstructure and Processing, 2007. **447**(1–2): pp. 277–284.

[45] Ottemer, X. and J.S. Colton, *Effects of aging on epoxy-based rapid tooling materials*. Rapid Prototyping Journal, 2002. **8**(4): pp. 215–223.

[46] Puebla, K., et al., *Effects of environmental conditions, aging, and build orientations on the mechanical properties of ASTM type I specimens manufactured via stereolithography*. Rapid Prototyping Journal, 2012. **18**(5): pp. 374–388.

[47] Mueller J, S.K. *The Effect of Build Orientation on the Mechanical Properties in Inkjet 3D Printing*. in *SFF Symposium*. 2015. Austin, TX.

[48] Bass, L., N.A. Meisel, and C.B. Williams, *Exploring variability of orientation and aging effects in material properties of multi-material jetting parts*. Rapid Prototyping Journal, 2016. **22**(5): pp. 826–834.

[49] Shen, A., et al., *UV-assisted direct write of polymer-bonded magnets*. Journal of Magnetism and Magnetic Materials, 2018. **462**: pp. 220–225.

[50] Rau, D.A., et al., *Ultraviolet-assisted direct ink write to additively manufacture all-aromatic polyimides*. ACS Applied Materials and Interfaces, 2018. **10**(41): pp. 34828–34833.

[51] Gonzalez, J.E.A., et al., *Hybrid direct ink write 3D printing of high-performance composite structures*. Rapid Prototyping Journal, 2023. **29**(4): pp. 828–836.

[52] Alrashdan, A., W.J. Wright, and E. Celik, *Light assisted hybrid direct write additive manufacturing of thermosets*. Proceedings of the ASME 2020 International Mechanical Engineering Congress and Exposition, IMECE 2020, Vol 2a, 2020.

[53] Rau, D.A., et al., *A dual-cure approach for the ultraviolet-assisted material extrusion of highly loaded opaque suspensions*. Additive Manufacturing, 2023. **72**: 103616.

[54] Esposito, G., et al., *Frontal polymerization for UV- and thermally initiated EPON 826 resin*. Polymer Engineering and Science, 2024. **64**(10): pp. 4760–4773.

[55] Ziaee, M., J.W. Johnson, and M. Yourdkhani, *3D printing of short-carbon-fiber-reinforced thermoset polymer composites via frontal polymerization*. ACS Applied Materials & Interfaces, 2022. **14**(14): pp. 16694–16702.

[56] Bhattacharjee, N., et al., *Desktop-stereolithography 3D-printing of a poly(dimethylsiloxane)-based material with Sylgard-184 properties*. Advanced Materials, 2018. **30**(22): 1800001.

[57] Patel, D.K., et al., *Highly stretchable and UV curable elastomers for digital light processing based 3D printing*. Advanced Materials, 2017. **29**(15): 1606000.

[58] Muth, J.T., et al., *Embedded 3D printing of strain sensors within highly stretchable elastomers*. Advanced Materials, 2014. **26**(36): pp. 6307–6312.

[59] Kuang, X., et al., *3D printing of highly stretchable, shape-memory, and self-healing elastomer toward novel 4D printing*. ACS Applied Materials and Interfaces, 2018. **10**(8): pp. 7381–7388.

[60] Chizari, K., et al., *3D printing of highly conductive nanocomposites for the functional optimization of liquid sensors*. Small, 2016. **12**(44): pp. 6076–6082.

[61] Postiglione, G., et al., *Conductive 3D microstructures by direct 3D printing of polymer/carbon nanotube nanocomposites via liquid deposition modeling*. Composites Part A – Applied Science and Manufacturing, 2015. **76**: pp. 110–114.

[62] Kwok, S., et al., *Electrically conductive filament for 3D-printed circuits and sensors*. Applied Materials Today, 2017. **9**: pp. 167–175.

[63] Al-Hariri, L.A.L., Branden; Nowotarski, Mesopotamia; Magi, James; Chambliss, Kaelynn; Venzel, Thaís; Delekar, Sagar; and S. and Acquah, *Carbon nanotubes and graphene as additives in 3D printing*. Carbon Nanotubes – Current Progress of their Polymer Composites, 2016: p. 1448.

[64] Ebrahimi, N.D. and Y.S. Ju, *Thermal conductivity of sintered copper samples prepared using 3D printing-compatible polymer composite filaments*. Additive Manufacturing, 2018. **24**: pp. 479–485.

[65] Laureto, J., et al., *Thermal properties of 3-D printed polylactic acid-metal composites*. Progress in Additive Manufacturing, 2017. **2**(1): pp. 57–71.

[66] Kim, K., et al., *3D optical printing of piezoelectric nanoparticle – polymer composite materials*. ACS Nano, 2014. **8**(10): pp. 9799–9806.

[67] Zhou, Y.Q., et al., *Embedding Carbon Dots in Superabsorbent Polymers for Additive Manufacturing*. Polymers, 2018. **10**(8) 921.

[68] Sweeney, C.B., et al., *Welding of 3D-printed carbon nanotube-polymer composites by locally induced microwave heating*. Science Advances, 2017. **3**(6): e1700262.

[69] N. Nawafleh, J.C., M. Aljaghtam, C. Oztan, E. Dauer, R. M. Gorguluarslan, T. Demir, E. Celik. *Additive manufacturing of kevlar reinforced epoxy composites*. in *IMECE 2019*. 2019. Salt Lake City, UT.

[70] Gray, R.W., D.G. Baird, and J.H. Bohn, *Effects of processing conditions on short TLCP fiber reinforced FDM parts*. Rapid Prototyping Journal, 1998. **4**(1): pp. 14–25.

[71] Zhong, W.H., et al., *Short fiber reinforced composites for fused deposition modeling*. Materials Science and Engineering A – Structural Materials Properties Microstructure and Processing, 2001. **301**(2): pp. 125–130.

[72] Rahimizadeh, A., et al., *Recycled glass fiber composites from wind turbine waste for 3D printing feedstock: Effects of Fiber Content and Interface on Mechanical Performance.* Materials, 2019. **12**(23): 3929.

[73] Ivey, M., et al., *Characterizing short-fiber-reinforced composites produced using additive manufacturing.* Advanced Manufacturing-Polymer & Composites Science, 2017. **3**(3): pp. 81–91.

[74] Tekinalp, H.L., et al., *Highly oriented carbon fiber-polymer composites via additive manufacturing.* Composites Science and Technology, 2014. **105**: pp. 144–150.

[75] E. Yasa, K.E., *A Review on the Additive Manufacturing of Fiber Reinforced Polymer Matrix Composites,* in *Solid Freeform Fabrication 2018.* 2018: Austin, TX.

[76] Pierson, H.A., et al., *Mechanical properties of printed epoxy-carbon fiber composites.* Experimental Mechanics, 2019. **59**(6): pp. 843–857.

[77] Ashby, M.F., *Materials Selection in Mechanical Design, 4th Ed.* 2011, Burlington, MA, USA: Elsevier.

[78] Van Hattum, F. and C. Bernardo, *A model to predict the strength of short fiber composites.* J Polymer composites, 1999. **20**(4): pp. 524–533.

[79] Fukuda, H. and T.W. Chou, *A probabilistic theory of the strength of short-fibre composites with variable fiber length and orientation.* Journal of Materials Science, 1982. **17**(4): pp. 1003–1011.

[80] Lauke, B. and S.Y. Fu, *Strength anisotropy of misaligned short-fibre-reinforced polymers.* Composites Science and Technology, 1999. **59**(5): pp. 699–708.

[81] Nakamoto, T. and S. Kojima, *Layered thin film micro parts reinforced with aligned short fibers in laser stereolithography by applying magnetic field.* Journal of Advanced Mechanical Design Systems and Manufacturing, 2012. **6**(6): pp. 849–858.

[82] Nakamoto, T., O. Kanehisa, and Y. Sugawa, *Whisker alignment in microparts using laser stereolithography with applied electric field.* Journal of Advanced Mechanical Design Systems and Manufacturing, 2013. **7**(6): pp. 888–902.

[83] Lewicki, J.P., et al., *3D-printing of meso-structurally ordered carbon fiber/polymer composites with unprecedented orthotropic physical properties.* Scientific Reports, 2017. **7**: 43401.

[84] Yang, D., et al., *A particle element approach for modelling the 3D printing process of fibre reinforced polymer composites.* Journal of Manufacturing and Materials Processing, 2017. **1**(1): 10.

[85] Oztan, C., et al., *Microstructure and mechanical properties of three dimensional-printed continuous fiber composites.* Journal of Composite Materials, 2019. **53**(2): pp. 271–280.

[86] Klift, Frank van der, Yoichiro Koga, Akira Todoroki, Masahito Ueda, Yoshiyasu Hirano and Ryosuke Matsuzaki. "3D Printing of Continuous Carbon Fibre Reinforced Thermo-Plastic (CFRTP) Tensile Test Specimens." Open Journal of Composite Materials 06 (2016): 18–27.

[87] Li, N.Y., Y.G. Li, and S.T. Liu, *Rapid prototyping of continuous carbon fiber reinforced polylactic acid composites by 3D printing.* Journal of Materials Processing Technology, 2016. **238**: pp. 218–225.

[88] Matsuzaki, R., et al., *Three-dimensional printing of continuous-fiber composites by in-nozzle impregnation.* Scientific Reports, 2016. **6**: 23058.

[89] Yang, C.C., et al., *3D printing for continuous fiber reinforced thermoplastic composites: mechanism and performance.* Rapid Prototyping Journal, 2017. **23**(1): pp. 209–215.

[90] Hao, W.F., et al., *Preparation and characterization of 3D printed continuous carbon fiber reinforced thermosetting composites.* Polymer Testing, 2018. **65**: pp. 29–34.

[91] Ming, Y.K., et al., *A novel route to fabricate high-performance 3D printed continuous fiber-reinforced thermosetting polymer composites.* Materials, 2019. **12**(9): 1369.

[92] *3D printing and additive manufacturing state of the industry, Wohlers Report 2019.* 2019.

[93] *All You Need to Know About Current 3D Metal Printing Technology.* 2019; Available from: https://engineeringcopywriter.com/3d-metal-printing-technology.

[94] *Engineering, Manufacturing, Metal Printing, Technology.* 2019; Available from: https://engineeringcopywriter.com/3d-metal-printing-technology.

[95] *Subcommittee F42.05 on Materials and Processes.* 05/25/2018; Available from: https://www.astm.org/COMMIT/SUBCOMMIT/F4205.htm.

[96] Sidambe, A.T., *Biocompatibility of advanced manufactured titanium implants – a review*. Materials, 2014. **7**(12): pp. 8168–8188.

[97] Dawes, J., R. Bowerman, and R. Trepleton, *Introduction to the additive manufacturing powder metallurgy supply chain exploring the production and supply of metal powders for AM processes*. Johnson Matthey Technology Review, 2015. **59**(3): pp. 243–256.

[98] Sun, P., et al., *Review of the methods for production of spherical Ti and Ti alloy powder*. JOM, 2017. **69**(10): pp. 1853–1860.

[99] Chen, G., et al., *A comparative study of Ti-6A1-4V powders for additive manufacturing by gas atomization, plasma rotating electrode process and plasma atomization*. Powder Technology, 2018. **333**: pp. 38–46.

[100] Hsu, T.I., et al., *Nitinol powders generate from plasma rotation electrode process provide clean powder for biomedical devices used with suitable size, spheroid surface and pure composition*. Scientific Reports, 2018. **8**: 13776.

[101] Herzog, D., et al., *Additive manufacturing of metals*. Acta Materialia, 2016. **117**: pp. 371–392.

[102] Skernivitz S., et al., *Industrial Issues in Additive Manufacturing*. 2017; Available from: https://www.digi talengineering247.com/article/industrial-issues-additive-manufacturing/.

[103] Langnau, L. *Will aluminum become the new hot material in metal additive manufacturing*. 2018; Available from: https://www.makepartsfast.com/will-aluminum-become-new-hot-new-material-in-metal-additive-manufacturing/.

[104] Chen, S.Y., Y. Tong, and P.K. Liaw, *Additive manufacturing of high-entropy alloys: a review*. Entropy, 2018. **20**(12): 937.

[105] Kenel, C., N.P.M. Casati, and D.C. Dunand, *3D ink-extrusion additive manufacturing of CoCrFeNi high-entropy alloy micro-lattices*. Nature Communications, 2019. **10**(1): p. 904.

[106] Zhang, H., et al., *Manufacturing and analysis of high-performance refractory high-entropy alloy via selective laser melting (SLM)*. Materials (Basel), 2019. **12**(5): 720.

[107] Patterson, A.E., S.L. Messimer, and P.A. Farrington, *Overhanging features and the SLM/DMLS residual stresses problem: review and future research need*. Technologies, 2017. **5**(2): 15.

[108] ASTM, *F3301, Standard for Additive Manufacturing – Post Processing Methods –Standard Specification for Thermal Post-Processing Metal Parts Made Via Powder Bed Fusion*.

[109] Nembach, E., *Particle Strengthening in Metals and Alloys*. 1997, New York: John Wiley & Sons.

[110] Gladman, T., *Precipitation Hardening in Metals*. Materials Science and Technology, 1999. **15**(1): pp. 30–36.

[111] ASTM, *A276, Standard Specification for Stainless Steel Bars and Shapes*.

[112] Mertens, A., et al., *Mechanical properties of alloy Ti-6Al-4V and of stainless steel 316L processed by selective laser melting: influence of out-of-equilibrium microstructures*. Powder Metallurgy, 2014. **57**(3): pp. 184–189.

[113] Yadollahi, A., et al., *Effects of process time interval and heat treatment on the mechanical and microstructural properties of direct laser deposited 316L stainless steel*. Materials Science and Engineering A – Structural Materials Properties Microstructure and Processing, 2015. **644**: pp. 171–183.

[114] Abd-Elghany, K. and D.L. Bourell, *Property evaluation of 304L stainless steel fabricated by selective laser melting*. Rapid Prototyping Journal, 2012. **18**(5): pp. 420–428.

[115] ASTM, *A564, Standard Specification for Hot-Rolled and Cold-Finished Age Hardening Stainless Steel Bars and Shapes*.

[116] T. Burkert, A.F. *The Effects of Heat Balance on the Void Formation Within Marage 300 Processed by Selective Laser Melting Symposium*, in *SFF*. 2015.

[117] Kempen, K., et al., *Microstructure and Mechanical Properties of Selective Laser Melted 18Ni-300 Steel*. Lasers in Manufacturing 2011: Proceedings of the Sixth International WLT Conference on Lasers in Manufacturing, Vol 12, Pt A, 2011. **12**: pp. 255–263.

[118] Mazumber, J., et al., *The direct metal deposition of H13 tool steel for 3-D components (vol 49, pg 55, 1997)*. JOM – Journal of the Minerals Metals & Materials Society, 1997. **49**(8): p. 8–8.

[119] DIN, *EN 1706, Aluminium and Aluminium Alloys – Castings – Chemical Composition and Mechanical Properties*. 2013.

[120] Prashanth, K.G., et al., *Microstructure and mechanical properties of Al-12Si produced by selective laser melting: Effect of heat treatment*. Materials Science and Engineering A – Structural Materials Properties Microstructure and Processing, 2014. **590**: pp. 153–160.

[121] D. Manfredi et al., Additive manufacturing of Al Alloys and Aluminium Matrix Composites (AMCs) in Light Met. Alloy. Appl. (2014), p. **15**.

[122] B.A. Fulcher, D.K.L., T.J. Watt, *Comparison of AlSi10Mg and Al 6061 Processed Through DMLS*, in *International Solid Freeform Fabrication Symposium*. 2014.

[123] Schmidtke, K., et al., *Process and Mechanical Properties: Applicability of a Scandium Modified Al-Alloy for Laser Additive Manufacturing*. Lasers in Manufacturing 2011: Proceedings of the Sixth International WLT Conference on Lasers in Manufacturing, Vol 12, Pt A, 2011. **12**: pp. 369–374.

[124] Bajoraitis R. et al., *Forming of Titanium and Titanium Alloys*, Vol. 14, ASM Metals Handbook, 1988.

[125] Attar, H., et al., *Manufacture by selective laser melting and mechanical behavior of commercially pure titanium*. Materials Science and Engineering A – Structural Materials Properties Microstructure and Processing, 2014. **593**: pp. 170–177.

[126] ASTM, *F1108-14, Standard Specification for Titanium-6Aluminum-4Vanadium Alloy Castings for Surgical Implants*. 2014.

[127] Donachie, M.J., *Titanium: A Technical Guide, 2nd edition*, A. International, Editor. 2000.

[128] Facchini, L., et al., *Ductility of a Ti-6Al-4V alloy produced by selective laser melting of prealloyed powders*. Rapid Prototyping Journal, 2010. **16**(6): pp. 450–459.

[129] Xu, W., et al., *Additive manufacturing of strong and ductile Ti-6Al-4V by selective laser melting via in situ martensite decomposition*. Acta Materialia, 2015. **85**: pp. 74–84.

[130] Carroll, B.E., T.A. Palmer, and A.M. Beese, *Anisotropic tensile behavior of Ti-6Al-4V components fabricated with directed energy deposition additive manufacturing*. Acta Materialia, 2015. **87**: pp. 309–320.

[131] Ren, H.S., et al., *Microstructural evolution and mechanical properties of laser melting deposited Ti-6.5Al-3.5Mo-1.5Zr-0.3Si titanium alloy*. Transactions of Nonferrous Metals Society of China, 2015. **25**(6): pp. 1856–1864.

[132] Riemer, A., et al., *On the fatigue crack growth behavior in 316L stainless steel manufactured by selective laser melting*. Engineering Fracture Mechanics, 2014. **120**: pp. 15–25.

[133] Sehrt, J.T., *Moglichkeiten und Grenzen bei der generativen Herstellung metallischer Bauteile durch das Strahlschmelzen*. 2010, University of Duisburg-Essen, Germany.

[134] Kammer, C., *Aluminium-Taschenbuch 1-Grundlagen und Werkstoffe*. 2012, Berlin: Beuth.

[135] Siddique, S., et al., *Influence of process-induced microstructure and imperfections on mechanical properties of AlSi12 processed by selective laser melting*. Journal of Materials Processing Technology, 2015. **221**: pp. 205–213.

[136] Buchbinder, D., *Selective Laser Melting von Aluminiumgusslegierungen*. 2013, RWTH Aachen.

[137] D. Greitemeier, K.S., V. Holzinger, C. Dalle Donne. *Additive layer manufacturing of Ti-6AL-4V and Scalmalloy*, in *27th ICAF Symposium*. 2013. Jerusalem.

[138] M. Peters, G.L., R.I. Jaffee, *Mechanical Properties of a Titanium Blading Alloy*, in *EPRI-report CS-2933*. 1983, Electric Power Research Institute: Palo Alto, CA, USA.

[139] Brandl, E., *Microstructural and Mechanical Properties of Additive Manufactured Titanium (Ti-6Al-4V) Using Wire*. 2010, TU Cottbus.

[140] Wycisk, E., et al., *Effects of defects in laser additive manufactured Ti-6Al-4V on fatigue properties*. 8th International Conference on Laser Assisted Net Shape Engineering (Lane 2014), 2014. **56**: pp. 371–378.

[141] Schwentenwein, M. and J. Homa, *Additive manufacturing of dense alumina ceramics*. International Journal of Applied Ceramic Technology, 2015. **12**(1): pp. 1–7.

[142] Khalyfa, A., et al., *Development of a new calcium phosphate powder-binder system for the 3D printing of patient specific implants*. Journal of Materials Science-Materials in Medicine, 2007. **18**(5): pp. 909–916.

[143] Friedel, T., et al., *Fabrication of polymer derived ceramic parts by selective laser curing*. Journal of the European Ceramic Society, 2005. **25**(2–3): pp. 193–197.

[144] Cesaretti, G., et al., *Building components for an outpost on the Lunar soil by means of a novel 3D printing technology*. Acta Astronautica, 2014. **93**: pp. 430–450.

[145] Fu, Z., et al., *Three-dimensional printing of SiSiC lattice truss structures*. Materials Science and Engineering A – Structural Materials Properties Microstructure and Processing, 2013. **560**: pp. 851–856.

[146] Moon, J., et al., *Fabrication of functionally graded reaction infiltrated SiC-Si composite by three-dimensional printing (3DP (TM)) process*. Materials Science and Engineering A – Structural Materials Properties Microstructure and Processing, 2001. **298**(1–2): pp. 110–119.

[147] Nan, B.Y., et al., *Three-dimensional printing of Ti3SiC2-based ceramics*. Journal of the American Ceramic Society, 2011. **94**(4): pp. 969–972.

[148] Sun, W., et al., *Freeform fabrication of Ti(3)SiC(2) powder-based structures Part I – Integrated fabrication process*. Journal of Materials Processing Technology, 2002. **127**(3): pp. 343–351.

[149] Yoo J, C.M., Khanuja S, Sachs E., *Structural Ceramic Components by 3D Printing*, in *Solid Freeform Fabrication Symposium*. 2003: Solid Freeform Fabrication Symposium.

[150] Fielding, G.A., A. Bandyopadhyay, and S. Bose, *Effects of silica and zinc oxide doping on mechanical and biological properties of 3D printed tricalcium phosphate tissue engineering scaffolds*. Dental Materials, 2012. **28**(2): pp. 113–122.

[151] Suwanprateeb, J., et al., *Mechanical and in vitro performance of apatite-wollastonite glass ceramic reinforced hydroxyapatite composite fabricated by 3D-printing*. Journal of Materials Science-Materials in Medicine, 2009. **20**(6): pp. 1281–1289.

[152] Deckers, J., et al., *Direct selective laser sintering/melting of high density alumina powder layers at elevated temperatures*. 8th International Conference on Laser Assisted Net Shape Engineering (Lane 2014), 2014. **56**: pp. 117–124.

[153] Shishkovsky, I., et al., *Alumina-zirconium ceramics synthesis by selective laser sintering/melting*. Applied Surface Science, 2007. **254**(4): pp. 966–970.

[154] Bertrand, P., et al., *Ceramic components manufacturing by selective laser sintering*. Applied Surface Science, 2007. **254**(4): pp. 989–992.

[155] Exner, H., et al., *Laser micro sintering – a new method to generate metal and ceramic parts of high resolution with sub-micrometer powder*. Virtual and Rapid Manufacturing, 2008: **3**(1): pp. 491–499.

[156] Schwarzer, E., et al., *Lithography-based ceramic manufacturing (LCM) – Viscosity and cleaning as two quality influencing steps in the process chain of printing green parts*. Journal of the European Ceramic Society, 2017. **37**(16): pp. 5329–5338.

[157] Altun, A.A.P., T.; Konegger, T.; Schwentenwein, M., *Dense, strong, and precise silicon nitride-based ceramic parts by lithography-based ceramic manufacturing*. Applied Sciences, 2020. **10**(3): p. 996.

[158] U. Scheithauer, E.S., G. Ganzer, A. Kornig, W. Becker, E. Reichelt, and A.H.t. M. Jahn, H. Richter, T. Moritz, *Additive manufacturing and strategic technologies in advanced ceramics*. Ceramic Transactions, 2015. **258**: pp. 31–41.

[159] Scheithauer, U., et al., *Additive manufacturing of ceramic heat exchanger: opportunities and limits of the lithography-based ceramic manufacturing (LCM)*. Journal of Materials Engineering and Performance, 2018. **27**(1): pp. 14–20.

[160] Lantada, A.D., et al., *Lithography-based ceramic manufacture (LCM) of auxetic structures: present capabilities and challenges*. Smart Materials and Structures, 2016. **25**(5): 054015.

[161] Rueschhoff, L., et al., *Additive manufacturing of dense ceramic parts via direct ink writing of aqueous alumina suspensions*. International Journal of Applied Ceramic Technology, 2016. **13**(5): pp. 821–830.

[162] Lewis, J.A., *Direct-write assembly of ceramics from colloidal inks*. Current Opinion in Solid State & Materials Science, 2002. **6**(3): pp. 245–250.

[163] Feilden, E., et al., *3D printing bioinspired ceramic composites*. Scientific Reports, 2017. **7**(1): p. 13759.

[164] Nguyen, V.L., *Structure-Property Relations of the Exoskeleton of the Ironclad Beetle*. 2017, Mississippi State University Starkville, Mississippi, USA.

[165] Jabbari, M., et al., *Ceramic tape casting: A review of current methods and trends with emphasis on rheological behaviour and flow analysis*. Materials Science and Engineering B-Advanced Functional Solid-State Materials, 2016. **212**: pp. 39–61.

[166] Klosterman, D.A., et al., *Development of a curved layer LOM process for monolithic ceramics and ceramic matrix composites*. Rapid Prototyping Journal, 1999. **5**(2): pp. 61–71.

[167] Pan, M.J., et al., *Optimizing the performance of telescoping actuators through rapid prototyping and finite element modeling*. Ceramic Materials and Multilayer Electronic Devices, 2003. **150**: pp. 53–62.

[168] C. Steidle, D.K., R. Chartoff, G. Graves, N. Osborne. *Automated Fabrication of Custom Bone Implants Using Rapid Prototyping*, in *44th International SAMPE Symposium and Exhibition*. 1999.

[169] Weisensel, L., et al., *Laminated object manufacturing (LOM) of SiSiC composites*. Advanced Engineering Materials, 2004. **6**(11): pp. 899–903.

[170] Griffin, E.A., D.R. Mumm, and D.B. Marshall, *Rapid prototyping of functional ceramic composites*. American Ceramic Society Bulletin, 1996. **75**(7): pp. 65–68.

[171] Danforth, S., *Fused deposition of ceramics: a new technique for the rapid fabrication of ceramic components*. Materials Technology, 1995. **10**: pp. 144–146.

[172] Safari, A., *Processing of advanced electroceramic components by fused deposition technique*. Ferroelectrics, 2001. **263**(1–4): pp. 1345–1354.

[173] Chen, Z.W., et al., *3D printing of ceramics: a review*. Journal of the European Ceramic Society, 2019. **39**(4): pp. 661–687.

[174] Colombo, P., et al., *Polymer-derived ceramics: 40 years of research and innovation in advanced ceramics*. Journal of the American Ceramic Society, 2010. **93**(7): pp. 1805–1837.

[175] Schelm, K., E.A. Morales, and M. Scheffler, *Mechanical and surface-chemical properties of polymer derived ceramic replica foams*. Materials, 2019. **12**(11): 1870.

[176] Eckel, Z.C., et al., *3D printing additive manufacturing of polymer-derived ceramics*. Science, 2016. **351**(6268): pp. 58–62.

[177] Fu, Y.L., et al., *Multiple metals doped polymer-derived SiOC ceramics for 3D printing*. Ceramics International, 2018. **44**(10): pp. 11030–11038.

[178] Zocca, A., et al., *Additive manufacturing of ceramics: issues, potentialities, and opportunities*. Journal of the American Ceramic Society, 2015. **98**(7): pp. 1983–2001.

[179] Wang, J.C., H. Dommati, and S.J. Hsieh, *Review of additive manufacturing methods for high-performance ceramic materials*. International Journal of Advanced Manufacturing Technology, 2019. **103**(5–8): pp. 2627–2647.

[180] Shahzad, K., et al., *Additive manufacturing of alumina parts by indirect selective laser sintering and post processing*. Journal of Materials Processing Technology, 2013. **213**(9): pp. 1484–1494.

[181] Zhang, Y., et al., *Al₂O₃ ceramics preparation by LOM (laminated object manufacturing)*. International Journal of Advanced Manufacturing Technology, 2001. **17**(7): pp. 531–534.

[182] Balla, V.K., S. Bose, and A. Bandyopadhyay, *Processing of bulk alumina ceramics using laser engineered net shaping*. International Journal of Applied Ceramic Technology, 2008. **5**(3): pp. 234–242.

[183] Wang, J.C. and H. Dommati, *Fabrication of zirconia ceramic parts by using solvent-based slurry stereolithography and sintering*. International Journal of Advanced Manufacturing Technology, 2018. **98**(5–8): pp. 1537–1546.

[184] Ebert, J., et al., *Direct inkjet printing of dental prostheses made of zirconia*. Journal of Dental Research, 2009. **88**(7): pp. 673–676.

[185] Corcione, C., et al., *Silica moulds built by stereolithography*. Journal of Materials Science, 2005. **40**(18): pp. 4899–4904.

[186] Klosterman, D.A., et al., *Structural composites via laminated object manufacturing (LOM)*. Solid Freeform Fabrication Proceedings, September 1996, 1996: pp. 105–115.

[187] A, T., *Freeze-Form Extrusion Fabrication of Boron Carbide*. 2015, Missouri University of Science and Technology.

[188] Vaidyanathan, R., et al., *The extrusion freeforming of functional ceramic prototypes*. Jom-Journal of the Minerals Metals & Materials Society, 2000. **52**(12): pp. 34–37.

[189] Kim, Y.K., et al., *Drop-on-demand inkjet-based cell printing with 30-mum nozzle diameter for cell-level accuracy*. Biomicrofluidics, 2016. **10**(6): p. 064110.

[190] Yi, H.G., H. Lee, and D.W. Cho, *3D printing of organs-on-chips*. Bioengineering (Basel), 2017. **4**(1): 10.

[191] Derakhshanfar, S., et al., *3D bioprinting for biomedical devices and tissue engineering: A review of recent trends and advances*. Bioactive Materials, 2018. **3**(2): pp. 144–156.

[192] Jaipan, P., A. Nguyen, and R.J. Narayan, *Gelatin-based hydrogels for biomedical applications*. MRS Communications, 2017. **7**(3): pp. 416–426.

[193] Choi, J.R., et al., *Recent advances in photo-crosslinkable hydrogels for biomedical applications*. Biotechniques, 2019. **66**(1): pp. 40–53.

[194] Heinrich, M.A., et al., *3D Bioprinting: from benches to translational applications*. Small, 2019. **15**(23): 1805510.

[195] Elieh-Ali-Komi, D. and M.R. Hamblin, *Chitin and chitosan: production and application of versatile biomedical nanomaterials*. International Journal of Advanced Research (Indore), 2016. **4**(3): pp. 411–427.

[196] N. Ashammakhi, S.A., C. Xua, H. Montazerian, H. Ko, R. Nasiri, N. Barros, A. Khademhosseini *Bioinks and bioprinting technologies to make heterogeneous and biomimetic tissue constructs*. Materials Today Bio, 2019. **1**: 100008.

[197] Dzobo, K., K.S.C.M. Motaung, and A. Adesida, *Recent trends in decellularized extracellular matrix bioinks for 3D printing: an updated review*. International Journal of Molecular Sciences, 2019. **20**(18): 4628.

[198] Ng, W.L., C.K. Chua, and Y.F. Shen, *Print me an organ! Why we are not there yet*. Progress in Polymer Science, 2019. **97**: 101145.

[199] Kumar, V.A., et al., *Tissue engineering of blood vessels: functional requirements, progress, and future challenges*. Cardiovascular Engineering Technology, 2011. **2**(3): pp. 137–148.

[200] Cubo, N., et al., *3D bioprinting of functional human skin: production and in vivo analysis*. Biofabrication, 2017. **9**(1): 015006.

[201] Orrhult, L.S., J. Sunden, and P. Gatenholm, *3D bioprinting of a human skin tissue model*. Tissue Engineering Part A, 2017. **23**: pp. S76–S76.

[202] Baltazar, T., et al., *Three dimensional bioprinting of a vascularized and perfusable skin graft using human keratinocytes, fibroblasts, pericytes, and endothelial cells*. Tissue Engineering Part A, (2020), **26**(5-6), 227–238.

[203] Michael, S., et al., *Tissue engineered skin substitutes created by laser-assisted bioprinting form skin-like structures in the dorsal skin fold chamber in mice* e57741. PLOS One, 2013. **8**(3).

[204] Zhang, L., J. Hu, and K.A. Athanasiou, *The role of tissue engineering in articular cartilage repair and regeneration*. Critical Reviews in Biomedical Engineering, 2009. **37**(1–2): pp. 1–57.

[205] Fellows, C.R., et al., *Adipose, bone marrow and synovial joint-derived mesenchymal stem cells for cartilage repair*. Frontiers in Genetics, 2016. **7**: 213.

[206] Armstrong, J.P.K., et al., *3D bioprinting using a templated porous bioink*. Advanced Healthcare Materials, 2016. **5**(14): pp. 1724–1730.

[207] Kikuchi, K. and K.D. Poss, *Cardiac regenerative capacity and mechanisms*. Annual Review of Cell and Developmental Biology, 2012. **28**: pp. 719–41.

[208] Noor, N., et al., *3D printing of personalized thick and perfusable cardiac patches and hearts*. Advanced Science, 2019. **6**(11): 1900344.

[209] Liu, J., et al., *Rapid 3D bioprinting of in vitro cardiac tissue models using human embryonic stem cell-derived cardiomyocytes*. Bioprinting, 2019. **13**: e00040.

[210] Beyersdorf, F., *Three-dimensional bioprinting: new horizon for cardiac surgery*. European Journal of Cardiothoracic Surgery, 2014. **46**(3): pp. 339–41.

[211] Jana, S. and A. Lerman, *Bioprinting a cardiac valve*. Biotechnology Advances, 2015. **33**(8): pp. 1503–1521.

[212] Alonzo, M., et al., *3D bioprinting of cardiac tissue and cardiac stem cell therapy*. Translational Research, 2019. **211**: pp. 64–83.

[213] Chandak, P., et al., *Patient-specific 3D printing: a novel technique for complex pediatric renal transplantation*. Annals of Surgery, 2019. **269**(2): pp. E18–E23.

[214] van den Berg, C.W., et al., *Renal subcapsular transplantation of PSC-derived kidney organoids induces neo-vasculogenesis and significant glomerular and tubular maturation in vivo*. Stem Cell Reports, 2018. **10**(3): pp. 751–765.

[215] Knowlton et al., *Bioprinting for cancer research*. Trends Biotechnol. 2015; **33**(9): 504–13.

[216] Oztan, C.Y., Nawafleh, N., Zhou, Y., Liyanage, P., Hettiarachchi S. D., Seven, E. S.,Leblanc, R. M., Ouhtit, A., Celik, E., *Recent advances on utilization of bioprinting for tumor modeling*. Bioprinting 2020. **18**: e00079.

[217] Ridky, T.W., et al., *Invasive three-dimensional organotypic neoplasia from multiple normal human epithelia*. Nature Medicine, 2010. **16**(12): pp. 1450.

[218] Ghosh, S., et al., *Three-dimensional culture of melanoma cells profoundly affects gene expression profile: a high density oligonucleotide array study*. Journal of Cellular Physiology, 2005. **204**(2): pp. 522–531.

[219] Kim, B.J., et al., *Cooperative roles of SDF-1α and EGF gradients on tumor cell migration revealed by a robust 3D microfluidic model*. Journal of Cellular Physiology, 2013. **8**(7): p. e68422.

[220] Zervantonakis, I.K., et al., *Three-dimensional microfluidic model for tumor cell intravasation and endothelial barrier function*. PNAS, 2012. **109**(34): pp. 13515–13520.

[221] Li, C.-L., et al., *Survival advantages of multicellular spheroids vs. monolayers of HepG2 cells in vitro*. Oncology Reports, 2008. **20**(6): pp. 1465–1471.

[222] Zhou, X., et al., *3D bioprinting a cell-laden bone matrix for breast cancer metastasis study*. ACS Applied Materials & Interfaces, 2016. **8**(44): pp. 30017–30026.

[223] Zhu, W., et al., *3D printed nanocomposite matrix for the study of breast cancer bone metastasis*. Nanomedicine: Nanotechnology, Biology and Medicine, 2016. **12**(1): pp. 69–79.

[224] Ling, K., et al., *Bioprinting-based high-throughput fabrication of three-dimensional MCF-7 human breast cancer cellular spheroids*. Engineering, 2015. **1**(2): pp. 269–274.

[225] Leonard, F. and B. Godin, *3D In Vitro Model for Breast Cancer Research Using Magnetic Levitation and Bioprinting Method*, in *Breast Cancer*. 2016, Springer. pp. 239–251.

[226] Grolman, J.M., et al., *Rapid 3D extrusion of synthetic tumor microenvironments*. Advanced Materials, 2015. **27**(37): pp. 5512–5517.

[227] Dai, X., et al., *Coaxial 3D bioprinting of self-assembled multicellular heterogeneous tumor fibers*. Scientific Reports, 2017. **7**(1): p. 1457.

[228] Dai, X., et al., *3D bioprinted glioma stem cells for brain tumor model and applications of drug susceptibility*. Biofabrication, 2016. **8**(4): p. 045005.

[229] van Pel, D.M., et al., *Modelling glioma invasion using 3D bioprinting and scaffold-free 3D culture*. Journal of Cell Communication and Signaling, 2018. **12**(4): pp. 723–730.

[230] Heinrich, M.A., et al., *3D-bioprinted mini-brain: a glioblastoma model to study cellular interactions and therapeutics*. Advanced materials, 2019: **31**(14): p. 1806590.

[231] Wang, X., et al., *Coaxial extrusion bioprinted shell-core hydrogel microfibers mimic glioma microenvironment and enhance the drug resistance of cancer cells*. Colloids and Surfaces B: Biointerfaces, 2018. **171**: pp. 291–299.

[232] Pang, Y., et al., *TGF-beta induced epithelial-mesenchymal transition in an advanced cervical tumor model by 3D printing*. Biofabrication, 2018. **10**(4): 044102.

[233] Zhao, Y., et al., *Three-dimensional printing of Hela cells for cervical tumor model in vitro*. Biofabrication, 2014. **6**(3): 035001.

[234] Huang, T.Q., et al., *3D printing of biomimetic microstructures for cancer cell migration*. Biomedical Microdevices, 2014. **16**(1): pp. 127–132.

[235] Zhang, B., et al., *3D bioprinting: an emerging technology full of opportunities and challenges*. Bio-Design and Manufacturing, 2018. **1**(1): pp. 2–13.

[236] Mehrotra, S., et al., *3D printing/bioprinting based tailoring of in vitro tissue models: Recent advances and challenges*. ACS Applied Bio Materials, 2019. **2**(4): 1385–1405.

[237] Wang, Y.J., et al., *Structural Design Optimization Using Isogeometric Analysis: A Comprehensive Review*. CMES – Computer Modeling in Engineering & Sciences, 2018. **117**(3): pp. 455–507.

[238] Deaton, J.D. and R.V. Grandhi, *A survey of structural and multidisciplinary continuum topology optimization: post 2000*. Structural and Multidisciplinary Optimization, 2014. **49**(1): pp. 1–38.

[239] Rozvany, G.I.N., *Grillages of maximum strength and maximum stiffness*. International Journal of Mechanical Sciences, 1972. **14**(10): pp. 651–+.

[240] Rozvany, G.I.N. and R.D. Hill, *Theory of optimal load transmission by flexure*. Advances in Applied Mechanics, 1976. **16**: pp. 183–308.

[241] Haber, R.B., C.S. Jog, and M.P. Bendsoe, *A new approach to variable-topology shape design using a constraint on perimeter*. Structural Optimization, 1996. **11**(1): pp. 1–12.

[242] Liu, J.K., et al., *Current and future trends in topology optimization for additive manufacturing*. Structural and Multidisciplinary Optimization, 2018. **57**(6): pp. 2457–2483.

[243] Huang, X.M., et al., *Sloping wall structure support generation for fused deposition modeling*. International Journal of Advanced Manufacturing Technology, 2009. **42**(11–12): pp. 1074–1081.

[244] Vanek, J., J.A.G. Galicia, and B. Benes, *Clever support: efficient support structure generation for digital fabrication*. Computer Graphics Forum, 2014. **33**(5): pp. 117–125.

[245] Dede, E.M., S.N. Joshi, and F. Zhou, *Topology optimization, additive layer manufacturing, and experimental testing of an air-cooled heat sink*. International Technical Conference and Exhibition on Packaging and Integration of Electronic and Photonic Microsystems, 2015, Vol 3, 2015.

[246] Christian M., M.E., and Lin P. T., *Heat Exchanger Design with Topology Optimization*, in *Heat Exchangers– Design, Experiment and Simulation*. 2017.

[247] Sokołowski J., Ż.A., *Topological Derivatives of Shape Functionals for Elasticity Systems*, in *Optimal Control of Complex Structures*, L.I. Hoffmann KH., Leugering G., Sprekels J., Tröltzsch F., Editor. 2001, Birkhäuser: Basel.

[248] Zhang, K., et al., *Reinitialization-free level set evolution via reaction diffusion*. IEEE Transactions on Image Processing, 2013. **22**(1): pp. 258–271.

[249] Luo, Z., et al., *Structural shape and topology optimization using a meshless Galerkin level set method*. International Journal for Numerical Methods in Engineering, 2012. **90**(3): pp. 369–389.

[250] Du W., B., Q., Zhang, B., *A novel method for additive/subtractive hybrid manufacturing of metallic parts*. Procedia Manufacturing, 2016. **5**: pp. 1018–1030.

[251] Kasperovich, G. and J. Hausmann, *Improvement of fatigue resistance and ductility of TiAl6V4 processed by selective laser melting*. Journal of Materials Processing Technology, 2015. **220**: pp. 202–214.

[252] Lopes, A.J., E. MacDonald, and R.B. Wicker, *Integrating stereolithography and direct print technologies for 3D structural electronics fabrication*. Rapid Prototyping Journal, 2012. **18**(2): pp. 129–143.

[253] Espalin, D., et al., *3D printing multifunctionality: structures with electronics.* International Journal of Advanced Manufacturing Technology, 2014. **72**(5–8): pp. 963–978.

[254] Perez, K.B. and C.B. Williams, *Design considerations for hybridizing additive manufacturing and direct write technologies.* Proceedings of the ASME International Design Engineering Technical Conferences and Computers and Information in Engineering Conference, 2014, Vol 4, 2014.

[255] Li, J., et al., *Hybrid additive manufacturing of 3D electronic systems.* Journal of Micromechanics and Microengineering, 2016. **26**(10): 105005.

[256] Jo, Y., et al., *3D polymer objects with electronic components interconnected via conformally printed electrodes.* Nanoscale, 2017. **9**(39): pp. 14798–14803.

[257] Giannitelli, S.M., et al., *Combined additive manufacturing approaches in tissue engineering.* Acta Biomaterialia, 2015. **24**: pp. 1–11.

[258] Bae, W.G., et al., *25th Anniversary Article: Scalable Multiscale Patterned Structures Inspired by Nature: the Role of Hierarchy.* Advanced Materials, 2014. **26**(5): pp. 675–699.

[259] Jensen, J., et al., *Surface-modified functionalized polycaprolactone scaffolds for bone repair: In vitro and in vivo experiments.* Journal of Biomedical Materials Research Part A, 2014. **102**(9): pp. 2993–3003.

[260] Puppi, D., et al., *Additive manufacturing of wet-spun polymeric scaffolds for bone tissue engineering.* Biomedical Microdevices, 2012. **14**(6): pp. 1115–1127.

[261] Mota, C., et al., *Additive manufacturing of star poly(epsilon-caprolactone) wet-spun scaffolds for bone tissue engineering applications.* Journal of Bioactive and Compatible Polymers, 2013. **28**(4): pp. 320–340.

[262] Hochleitner, G., et al., *Additive manufacturing of scaffolds with sub-micron filaments via melt electrospinning writing.* Biofabrication, 2015. **7**(3): 035002.

[263] Li, J.L., et al., *Fabrication of three-dimensional porous scaffolds with controlled filament orientation and large pore size via an improved E-jetting technique.* Journal of Biomedical Materials Research Part B – Applied Biomaterials, 2014. **102**(4): pp. 651–658.

[264] A. El-Desouky, M.C., M.A. Andre, P.M. Bardet, S. LeBlanc, *Rapid processing and assembly of semiconductor thermoelectric materials for energy conversion devices.* Materials Letters, 2016. **185**: pp. 598–602.

[265] Aljaghtham, M. and E. Celik, *Effect of Leg Topologies on Thermal Reliability of Thermoelectric Generators Systems.* Proceedings of ASME 2021 International Mechanical Engineering Congress and Exposition (IMECE2021), Vol 8a, 2021.

[266] Aljaghtham, M. and E. Celik, *Design of cascade thermoelectric generation systems with improved thermal reliability.* Energy, 2022. **243**: 123032.

[267] Aljaghtham, M. and E. Celik, *Numerical analysis of energy conversion efficiency and thermal reliability of novel, unileg segmented thermoelectric generation systems.* International Journal of Energy Research, 2021. **45**(6): pp. 8810–8823.

[268] Aljaghtham, M. and E. Celik, *Energy conversion and thermal reliability of thermoelectric materials in unileg annular configuration.* Materials Letters, 2021. **300**: 130192.

[269] He, M., et al., *3D printing fabrication of amorphous thermoelectric materials with ultralow thermal conductivity.* Small, 2015. **11**(44): pp. 5889–94.

[270] Oztan, C., et al., *Additive manufacturing of thermoelectric materials via fused filament fabrication.* Applied Materials Today, 2019. **15**: pp. 77–82.

[271] Kim, F., et al., *3D printing of shape-conformable thermoelectric materials using all-inorganic Bi2Te3-based inks.* Nature Energy, 2018. **3**(4): pp. 301–309.

[272] El-Desouky, A., et al., *Rapid processing and assembly of semiconductor thermoelectric materials for energy conversion devices.* Materials Letters, 2016. **185**: pp. 598–602.

[273] Mao, Y., et al., *Non-equilibrium synthesis and characterization of n-type Bi2Te2.7Se03 thermoelectric material prepared by rapid laser melting and solidification.* RSC Advances, 2017. **7**(35): pp. 21439–21445.

[274] Qiu, J.H., et al., *3D Printing of highly textured bulk thermoelectric materials: mechanically robust BiSbTe alloys with superior performance.* Energy & Environmental Science, 2019. **12**(10): pp. 3106–3117.

[275] Zhang, H.D., et al., *Laser additive manufacturing of powdered bismuth telluride.* Journal of Materials Research, 2018. **33**(23): pp. 4031–4039.

[276] Shi, J.X., et al., *3D printing fabrication of porous bismuth antimony telluride and study of the thermoelectric properties.* Journal of Manufacturing Processes, 2019. **37**: pp. 370–375.

[277] Ferhat, S., et al., *Organic thermoelectric devices based on a stable n-type nanocomposite printed on paper.* Sustainable Energy & Fuels, 2018. **2**(1): pp. 199–208.

[278] Yan, Y.G., et al., *Thermoelectric properties of n-type ZrNiSn prepared by rapid non-equilibrium laser processing.* RSC Advances, 2018. **8**(28): pp. 15796–15803.

[279] Choo, S., et al., *Cu Se-based thermoelectric cellular architectures for efficient and durable power generation.* Nature Communications, 2021. **12**(1): 3550.

[280] Gustinvil, R., et al., *Enhancing conversion efficiency of direct ink write printed copper(I) sulfide thermoelectrics via sulfur infusion process.* Machines, 2023. **11**(9): 881.

[281] S., T. *The Emergence of "4D Printing".* www.ted.com/talks/skylar_tibbits_the_emergence_of_4d_printing 2013.

[282] Tibbits, S., et al., *4D printing and universal transformation.* Acadia 2014: Design Agency, 2014: pp. 539–548.

[283] Tibbits, S., *4D printing: multi-material shape change.* Architectural Design, 2014. **84**(1): pp. 116–121.

[284] Gladman, A.S., et al., *Biomimetic 4D printing.* Nature Materials, 2016. **15**(4): pp. 413–+.

[285] Wu, Z.L., et al., *Three-dimensional shape transformations of hydrogel sheets induced by small-scale modulation of internal stresses.* Nature Communications, 2013. **4**: 1586.

[286] Naficy, S., et al., *4D printing of reversible shape morphing hydrogel structures.* Macromolecular Materials and Engineering, 2017. **302**(1): 1600212.

[287] Kuang, X., et al., *Advances in 4D Printing: Materials and Applications.* Advanced Functional Materials, 2019. **29**(2): 1805290.

[288] Zarek, M., et al., *3D printing of shape memory polymers for flexible electronic devices.* Advanced Materials, 2016. **28**(22): pp. 4449.

[289] Mao, Y.Q., et al., *Sequential self-folding structures by 3D printed digital shape memory polymers.* Scientific Reports, 2015. **5**: 13616.

[290] Gonzalez-Henriquez, C.M., M.A. Sarabia-Vallejos, and J. Rodriguez-Hernandez, *Polymers for additive manufacturing and 4D-printing: materials, methodologies, and biomedical applications.* Progress in Polymer Science, 2019. **94**: pp. 57–116.

[291] Roach, D.J., et al., *Novel ink for ambient condition printing of liquid crystal elastomers for 4D printing.* Smart Materials and Structures, 2018. **27**(12): 125011.

[292] Kotikian, A., et al., *3D Printing of Liquid Crystal Elastomeric Actuators with Spatially Programed Nematic Order.* Advanced Materials, 2018. **30**(10): 1706164.

[293] Campbell, T.A., S. Tibbits, and B. Garrett, *The Programmable World.* Scientific American, 2014. **311**(5): pp. 60–65.

[294] Ge, Q., et al., *Multimaterial 4D printing with tailorable shape memory polymers.* Scientific Reports, 2016. **6**: 31110.

[295] Zhang, Z.Z., K.G. Demir, and G.X. Gu, *Developments in 4D-printing: a review on current smart materials, technologies, and applications.* International Journal of Smart and Nano Materials, 2019. **10**(3): pp. 205–224.

[296] Wu, C., et al., *Topology optimisation for design and additive manufacturing of functionally graded lattice structures using derivative-aware machine learning algorithms.* Additive Manufacturing, 2023. **78**: 103833.

[297] Gu, G.X., et al., *Bioinspired hierarchical composite design using machine learning: simulation, additive manufacturing, and experiment.* Materials Horizons, 2018. **5**(5): pp. 939–945.

[298] Wang, R.J., et al., *Optimizing process parameters for selective laser sintering based on neural network and genetic algorithm*. International Journal of Advanced Manufacturing Technology, 2009. **42**(11–12): pp. 1035–1042.

[299] Lee, S.H., et al., *A neural network approach to the modelling and analysis of stereolithography processes*. Proceedings of the Institution of Mechanical Engineers Part B – Journal of Engineering Manufacture, 2001. **215**(12): pp. 1719–1733.

[300] Qin, J., et al., *Research and application of machine learning for additive manufacturing*. Additive Manufacturing, 2022. **52**: 102691.

[301] Nguyen, N.V., et al., *Semi-supervised machine learning of optical monitoring data for anomaly detection in laser powder bed fusion*. Virtual and Physical Prototyping, 2023. **18**(1): e2129396.

[302] Wright, W.J., et al., *In-situ optimization of thermoset composite additive manufacturing via deep learning and computer vision*. Additive Manufacturing, 2022. **58**: 102985.

Index

https://doi.org/10.1515/9781501520242-010